WEB
On-The-Go

*A Preview of the Way of Living
in a Wireless Web World*

BALA SANKAR

WEB ON-THE-GO
A Preview of the Way of Living in a Wireless Web World
Copyright © 2022   Bala Sankar

Library of Congress Control Number:	2022936921
Paperback:	978-1-958169-23-0
eBook:	978-1-958169-25-4

Printed in the United States of America

Contents

Preface 1

Vision 11

Overview 21

Zoo 41

Stranded In The Woods 51

Shipping Containers 63

Intelligent Vehicles 77

Educational Institutions 89

Security 99

Shopping 105

Business Enterprises 111

City Utilities 123

Nature 135

The Beginning 145

References 151

PREFACE

Economist vs. Innovator

Peter Ferdinand Drucker's first book was entitled The End of Economic Man, published in 1939. In that book, Drucker wrote the following paragraph:

"It is what Joseph Schumpeter (1883 – 1950) did when he identified the "innovator" as the social force that turns economies upside down; the innovator does not behave economically, does not try to optimize, is not motivated by economic rationale – innovator is a social phenomenon. It is what this book tries to do."

If you watch a football, soccer or basketball game, you see many different players enter the arena. During the game, some players might try brand new moves, while others might introduce

enhancements to the existing strategy. All players take risks; they play hard and try to win every game.

When you watch the game on television, you also may see a few well-dressed people sitting around a table in a comfortable room, making commentaries about the game. You might hear these commentators offer their ideas, such as: "Michael Jordan should have gone a few steps in front and then tried to shoot the ball. That would have been the perfect shot. Had he done that, the Chicago Bulls would have won the game."

People listening to these sports commentaries may agree with the analysis. Listeners might believe a certain commentator to be a genius when it comes to sports strategy, but we don't for a moment presume that these commentators should leave their comfortable chairs and take Michael Jordan's place on the team. That would be ridiculous.

We know that the commentators are not the players. Players keep their bodies and minds fit to play the game. Players go to the game with new strategies or enhanced existing strategies, every time they enter the field. Players take risks and know the challenges. They don't succeed all the time, but they still get up and play the next game with a strong determination to win.

Economists are not players. In this analogy, economists are very much like sports commentators. They can only make comments on the players' performances, and of course they are only able to comment on the past performances of the players. Their present statement is always based on what has already happened. Economists have no idea about present activity and potential future results. They must live in the past.

And yet economists are so incredibly influential. Their statements stay in the public mind as powerful indicators of what is actually happening. When economists make a statement such as, "The economy is in recession," people panic. The stock market plummets. When economists make a statement such as, "The economy is in depression," people jump out of windows.

Even the most innovative organizations in the world are now making statements like, "We are not immune from the economic depression," and "We view the situation as dire." CEOs are making very grim public remarks about the economy.

But the truth is, these economists are much like sports commentators. They are not players in the actual game. They are talking about the past.

Then who are the players who are actually making things happen now?

The players are the innovators. Innovators build the future now. Innovation creates wealth. The innovator is the social force that turns economies upside down. The innovator does not behave economically, does not try to optimize, and is not motivated by economic rationale. The innovator is a social phenomenon. Only innovation creates wealth, not economists' assessments such as "recession" and "depression."

Peter F. Drucker tried to do this in his first book, The End of Economic Man. He tried to free society from the clutches of economists and their fearful statements such as "recession" and "depression." In almost all of his books, talks, and lectures throughout his entire life, Drucker tried to shift the focus away from these scary economic statements, and instead replace them with words such as: "innovation" and "innovative opportunities."

Even today, our focus remains on the present economic statements being made, which are based on past reports from various sectors of the economy such as housing, jobs and GDP. If we remain stuck in the past, completely occupied with present economic statements, our future will be in recession and depression. The truth is, the economy is not in recession; the economy is not in depression; it is innovation that is suffering. Right now, innovation is in recession; innovation is in depression. Organizations have ceased all innovation; organizations have stopped searching for opportunities for innovation. The innovators are holding back. The economists are at the forefront, continuing to scare people with their statements.

We do not need more economists; we do need more innovators; innovators are the social force that will turn economies around. The intention of this book is to shift our focus from the economists to the innovators and innovation, in particular in the field of wireless web and emerging mobile technology.

Innovation and Wealth

According to Drucker, innovation is not a technical term belonging strictly to the field of science or technology. Innovation is, rather, an economic and social term, pointing toward a shift in the overall economic or social environment of the day. This shift could include changes in people's behavior as producers, consumers, students, teachers, etc...

He goes on to explain that rather than creating new knowledge, innovation actually creates new wealth or new potential for action. Innovation could be thought of as the exploitation of new ideas. If this is true, then most innovation will stem from the places that are already rich in manpower, brainpower and money, namely, existing businesses, schools and public service institutions. Drucker believed that innovation was absolutely pivotal if a society hoped to maintain and create high-quality jobs and keep successful businesses afloat. Obviously then, our challenge now is to create a society where innovation comes first, in businesses, schools and governments. Governments must play a key role in this shift toward innovation, but they can't do it alone.

If we continue thinking in these terms, then in essence, the economic stimulus package initiated by the administration in 2009 was meant simply to jump-start our society's innovation. The U.S. Congress put aside $4.7 billion in order to establish a Broadband Technology Opportunities Program. The

BTOP program offered awards to eligible entities in order to develop and expand broadband services to unserved and underserved areas. Awards were also given out to improve access to broadband by public safety agencies; $250 million was set aside for innovative programs that intended to adopt sustainable broadband services; another $200 million was set aside to upgrade technology and capacity at public computing centers, including community colleges and public libraries; $10 million was transferred to the Office of Inspector General for the purposes of BTOP audits and oversight; up to $350 million of the BTOP funding was designated for the development and maintenance of statewide broadband inventory maps.

As we move into this next era, and new money for broadband services becomes available, it is extremely important to remember the importance of innovation. Peter F. Drucker stressed the importance of certain principles of innovation that are of the utmost relevance today:

1. Purposeful, systematic innovation begins with the analysis of the sources of opportunities.
2. To be effective, an innovation has to be simple, and it has to be focused.
3. Effective innovations start small.
4. The successful innovation aims from the beginning to become the standard setter, to determine the direction of a new technology or a new industry, to create the business that is – and remains – ahead of the pack.
5. Innovation is work rather than genius. It requires knowledge. It requires ingenuity. And it requires focus.

Innovative Opportunities

Organizations must execute conscious and purposeful searches for innovative opportunities. This book is meant to be used as a guide for organizations to help them pinpoint some of these opportunities. It includes practical examples that will encourage any organization to identify key opportunities for innovation. Once an organization has identified an attractive opportunity, then the company will still need to take a leap of faith and imagination in order to arrive at the right response. First, organizations must know what innovation means; as I mentioned earlier, innovation is not science or technology, but value. Innovation does not take place within an organization, but rather is a change outside the organization. The measure of innovation could very well be the impact of the innovation on the environment. Innovation in a business enterprise must, therefore, always be market-focused.

Bell Labs invented transistor technology. But the scientists there saw innovation from a strictly technological vantage point. They failed to understand the value of their innovation. Those scientists told Sony's Akio Morita, who approached them with a request to use transistor technology, that the transistor was only good for use in the hearing aid. Had they asked Sony's Akio Morita about his plans for the use of the transistor, perhaps Sony might have shared his vision about the mobile radio. Who knows? Maybe Sony and Bell Labs would have come together to manufacture the mobile radio. Had this happened, those scientists at Bell Labs would have earned a great deal of money off their transistor innovation. But the truth is, the Bell Labs scientists didn't make the money. Sony did.

This book helps engineers, scientists and technologists pinpoint specific potential technological applications, and at the same time challenge them to actually build these applications. If innovation is successful, it is likely to turn into a major product or process, a major new business, and a major market.

To manage innovation, a manager does not need to be a technologist. Indeed, a first-rate technologist is rarely good at managing innovation. Technologists are so deeply engrossed in their specialty that they rarely see outside development.

Similarly, an innovative manager need not be an economist. Economists can concern themselves with the impact of innovation only after the product has become a massive force.

The innovative manager must be able to anticipate vulnerabilities and opportunities. This person needs to study innovation and learn its dynamics, its patterns, and its predictability. The innovative manager converts impractical, half-baked, and wild ideas into concrete, innovative reality. In the innovative organization, top management listens to ideas and takes them seriously.

Top management in the innovative organization knows that new ideas are often impractical; they also know that it takes a great many silly ideas to give birth to a fantastic one. In the early stages, there is no way to differentiate the silly ideas from the ones that will eventually change our lives; both look equally impossible or equally brilliant. Senior executives in the innovative organization should encourage ideas and ask how they can make ideas practical, realistic and effective. Senior executives should organize themselves so that they can think through even the wildest and apparently silliest idea for something new, so that the feasibility of the idea can be appraised. Top management in the innovative organization is the major driver for innovation. They can use the ideas in this book to stimulate vision and then work to make these new ideas a prominent concern of the entire organization.

Resistance to change is grounded in ignorance and in fear of the unknown. However, if change is seen as an opportunity, then there will be no fear. One way to organize innovative units within a large business might well be to group them together into one group. This group shall report to one member of top management, who has no other function, but to guide, help, advise, review, and direct the innovating team. In this wireless web innovation, the group may have software developers, wireless infrastructure professionals, marketing professionals and government liaison people. The team must consist of cross-functional professionals who together can take the wild, silly ideas and come up with a new innovative product or process.

For innovative strategy, the device must be "new and different." The basis of innovative strategy must include the elimination of the old technology that no longer serves. Innovating organizations should not spend time or resources defending yesterday. Indeed, systematic abandonment of yesterday can free resources, especially the most important resource of all -- capable people, for work on the new technology. Unwillingness to do this could very well be the greatest blockade to innovation in existing large businesses.

The innovative organization resists stagnation rather than change. This book helps the organization to be innovative. This book itself is an innovator. An "innovator" is a social force that turns economies around. The innovator does not behave economically, does not try to optimize, and is not motivated by economic rationale – the innovator is a social phenomenon.

Communication and Transportation

Human beings need two kinds of freedom:
1. The ability to know, read, speak, write, watch, listen and express. This need is fulfilled by communication tools.
2. The ability to physically move around anywhere on planet Earth. This need is fulfilled by transportation facilities.

In this book, human beings are granted both freedoms through the use of functioning systems of communication and transportation.

Audiences

Scientists, engineers and technologists have an abundance of knowledge about the latest technology as well as the appropriate skill sets, but little idea about customer needs. This book helps to identify customer needs so that scientists, engineers and technologists can build applications to meet those needs.

The engineering team may suggest this book to the marketing team of an organization. The marketing team would then research the customer needs and requirements in complete detail and provide the engineers with a challenge to their knowledge and skill sets. Engineers love challenges.

The marketing team may suggest this book to government agencies and government executives, policy makers and politicians, because this book will help them make decisions, such as providing wireless infrastructure throughout the city or state.

Finally, this book will help the general public to prepare for the future. It sets high expectations for Web on-the-go infrastructure. Here, we do not discuss the technology or the science behind the wireless Web; rather, we highlight the value of innovation, and of Web on-the-go.

Motivation for the Book

Around the end of 2007, the industry was speculating about the Google phone. But, instead of making a phone, Google released the open platform called Android, designed for mobile devices. In order to promote Android, Google announced the developer challenge. They asked for people to submit proposals for applications based on the Android platform.

I was inspired by the challenge and began thinking about what kind of applications could be built. I came up with many innovative ideas. Instead of building just one application, I went on to prepare a document with illustrations of the applications that I had begun to envision. That document is the basis and inspiration for this book.

Traditionally, innovators like Arthur C. Clarke expressed their new ideas by publishing science fiction novels. This book is not science fiction, but it is filled with unique innovations that do not yet exist.

Acknowledgments

It took me some time to switch gears from software development to actual publication of a book. I was looking for professional support in creative writing, editing, formatting and the creation of illustrations. Step by step, I built a team of experts. Special care has been taken to build a team with social and political knowledge, as well as expertise in their given fields.

This global team includes: Meg Hamill in Northern California; Linda Jay Geldens in Kentfield, California; and Pippa Cornell, who lives in the UK. All of these individuals helped to make this project a success. I want to thank and appreciate Meg Hamill for her creative writing expertise. I want to thank Linda Jay Geldens for her expertise in copyediting and proofreading. I want to thank Pippa Cornell for creating the fantastic illustrations that accompany these stories.

My Family

While I was consulting at Google, more than their technology, their kitchen and food inspired me a great deal. Google is like a Willy Wonka Chocolate Factory for software professionals. Later, I thought of entering the food business. But my wife Sudha said that business was her future plan and asked me to look for something else.

Having worked in the computer industry for nearly 25 years, and having witnessed the technology growth that accompanied it, I was convinced that now was time to take this technology to the people and society at large. So I told my wife that I was going to work on publishing a book on innovation. I would like to thank my wife for offering me total support on this new venture.

My daughter Luxwin had been suggesting that I buy the Google G1 phone. However I kept postponing this decision. One day, my daughter accidentally stepped on my old phone's charging wire; the phone fell and broke into two pieces. That accident was an innovative opportunity for me. I then bought the Google G1 from T-Mobile. Having the actual working phone on helped me to come up with the ideas in this book.

In 2008, my cell phone bill suddenly doubled. After investigation, I found that my son Ashwin had subscribed for a joke-a-day service, after watching a commercial on television. For a few lines of text each day, the cell phone companies were charging me an exorbitant amount of money. This didn't seem just to me. I decided that it was time for the systematic abandonment of the cell phone. Now I have Google G1 phone service from T-Mobile. My son has total freedom and can browse the Internet on-the-go using this new innovation. He also downloads tons of applications, from games to creative arts.

Open Mind and Open Ideas

With so many innovative ideas, I had two choices. One choice was to be greedy and to close my mind. That way, I might have come up with one or two ideas, then patented them to make millions of dollars. The other choice was to open up my mind, be responsible and share my ideas. When I do that, then millions of ideas seem to fly out of my open mind. By publishing those ideas in this book, I decided I could still make millions of dollars. Obviously, the second choice won out!

The founding fathers of America didn't patent the Constitution. Peter Drucker shared a myriad of innovative ideas in his books. Arthur C. Clarke didn't feel the need to patent his ideas.

Let me follow in the footsteps of these visionaries and dedicate this book to all of the innovators who have helped, and are helping to, build this great world.

VISION

"People only see what they are prepared to see," **Ralph Waldo Emerson (1803 – 1882)**

Human beings have created, and are living within, a complicated web of politics and invention, scientific discovery, and architectural wonder. Every time we step into a building, turn on the car, phone a friend, or cast a fishing line, we are interacting with the physical manifestation of someone else's dream. Every tool that we use, every human-made construction, began as vision, imagination, and pure potential within the human brain. Vision is, however, simply one aspect of the equation. Visions must be supported by innovation, conceptualization, and hard work, if they are ever to become tangible systems and objects in our human society.

Today we live in a time when technological advances are happening so quickly that it truly is difficult to keep up. It is almost unfathomable to imagine what the future will hold, what visions are being kindled right now in the minds of ordinary citizens, scientists, engineers, teachers, and poets all over the world.

A recurrent theme throughout history seems to be that paradigm shifts in human consciousness are extremely difficult in the moment, yet also essential for progress. This is true in the technological realm as much as in the philosophical realm. Being able to comprehend the potential of new technology has always been challenging. Many engineers and inventors could never have imagined the ways in which their invention would be practically applied in the world. For example, when Guglielmo Marconi invented the radio, he assumed it would be used for person-to-person communication. Today, radio broadcasting allows people living in the far corners of our Earth to hear the same news. Surprisingly, Alexander

Graham Bell invented the telephone while working to assist deaf people; he thought his technology would be used to broadcast concerts.

It can generally be said that scientists and engineers are somewhat limited in pinpointing the potential applications for their new technology. Is it really even their job? At some point, however, someone must step in who intimately understands developments on the technological front and can also envision the potential future use of that technology. It is up to this person to find a way to connect existing technology with a vision for how to apply it in new and as-yet-undiscovered ways.

A similar situation is happening right now in the field of mobile Internet technology. Many of us have easy access at home to computers and the Internet, which is loaded with incredible information. However, the Internet is oftentimes not available when we are traveling or on the road. Today we truly live in a global society. Many jobs require us to travel great distances and spend extended amounts of time away from our families and the comforts of domestic life. Families are split apart by the demands of jobs, and many of us find ourselves traveling thousands of miles just to come home from our career responsibilities. Even if we live close to our job and our family, we inevitably spend time each day away from our computers and easy access to the Internet.

With so much mobility in our society, with so many of us traveling more and more, I would suggest that it is indeed the right time to make mobile Internet technology not only commonplace, but easy to use, affordable, and beneficial to our communities. This book will outline my thoughts and ideas on why mobile Internet technology is important, and how to go about putting these systems in place.

I will explain to the best of my ability the evolving technology in Mobile Internet Devices, Internet Everywhere, and Wireless Web. Once I have laid out the details of how this technology works, I will demonstrate the potential, as I envision it, for Mobile web applications and services. I will do this by showing how this technology can be applied in a variety of specific life situations. My hope is that through writing this book, I will be able to bridge the gap between existing mobile Internet technology and practical applications for this technology that do not yet exist.

Let's begin by taking a moment to look back over our love affair with technology and invention. Our history is a rich patchwork collaboration among great visionaries, inventors, and engineers.

Each generation does its part to further technological advancements and quality of life for the next. Countless times in our past, we have found ourselves with great technology available, yet without a vision for how to apply that in the world. Oftentimes it takes the unplanned collision of one person's technological breakthrough with another person's unrelated vision in order to make these giant leaps in human technological evolution. In the pages that follow I will describe some of the key characters and companies involved in two very important breakthroughs in human history: putting a man on the moon, and the creation of the transistor radio.

One could argue that it is never a single person, or a single moment in history, that allows a particular vision to come to fruition. Rather, each technological advancement follows thousands of years of human ingenuity and progress. Before 1969 there was a long and winding chain of fantastic human minds that led us to put two men on the moon. I will not attempt to start at the beginning of this chain, as that might prove to be an impossible task; instead, I'll start with a man widely referred to as "the greatest genius ever."

Leonardo da Vinci

One of the greatest visionaries ever known was Leonardo da Vinci, born in Vinci, Italy, near Florence. In his life, Leonardo cultivated a new way of thinking about machines. He developed the idea that if he could gain an understanding for how individual parts of machines functioned, he could then change these parts and improve them, as well as combine them differently to create entirely new machines. While he was alive, da Vinci was recognized for his engineering skills and as an inventor. As a scientist, he furthered the knowledge base in the fields of anatomy, civil engineering, optics, and hydrodynamics. Because Leonardo was also an artist (perhaps his most famous painting is the Mona Lisa) he was able to illustrate his visions on paper with extraordinary clarity. His journals include plans for musical instruments, hydraulic pumps, reversible crank mechanisms, finned mortar shells, and a steam cannon, among others. One concept that appeared and reappeared in da Vinci's journals was that of human flight.

When Leonardo was an infant, he saw a hawk hovering above his cradle. In his own words: "I was in my cradle and a great hawk flew down to me. It opened my mouth with its tail and its feathers struck me several times inside my lips. That bird seems to me now to have pointed me to my destiny." Leonardo felt that he was destined to fly, and he spent a great deal of time and energy trying to do so. Leonardo wrote a text entitled

Codex on the Flight of Birds, where he recorded the first known scientific observations about flight. Leonardo sketched illustrations for a glider, a parachute, helical wings, beating wings, bats' wings, and a helicopter. The glider plans that he conceived turned into the world's first plane design capable of flight, known as the da Vinci flying machine. Gliders today are closely related to that machine. Whether or not da Vinci flew his own glider in his lifetime is unknown. Just a few years ago, however, one of Leonardo's glider designs were recreated, using materials da Vinci would have had available to him. The glider flew for a longer distance and at a higher altitude than the famous flight of the Wright brothers.

Today there is growing interest in replicating da Vinci's models and illustrations. In da Vinci's day, many of the materials necessary to manifest his multitudes of visions were simply not available, whereas nowadays they are. Oftentimes, however, da Vinci's models are not realistic or complete, which makes replication confusing and does not always lead to a functional creation. However, Leonardo da Vinci was not afraid to envision, contemplate, and create what most of the people around him deemed utterly impossible. This fearlessness has led da Vinci to be widely referred to today as "the greatest genius ever."

Leonardo da Vinci has been called a man who "saw the future." He made substantial contributions to both the scientific and artistic worlds, and left a legacy that continues to be of the utmost importance today. One of his most important contributions was his vision of a human being in flight. In the following pages, I will contemplate how this initial dream of flight has led our species towards some of the most unimaginable accomplishments, including a man walking on the moon.

Arthur C. Clarke

Fittingly, let's fast-forward 500 years, and turn our attention to a man who has been referred to as "the Leonardo da Vinci of our time." Arthur C. Clarke made great strides in both the scientific and literary communities, and, like da Vinci, he has been celebrated as a man who "saw the future." Clarke played an instrumental role in some of the greatest technological strides in history, and, like da Vinci, used his artistic abilities to do so. Clarke left us with an impressively high stack of celebrated books.

Arthur C. Clarke was a British science fiction author, futurist, and inventor, most well known for the novel The Sentinel, adapted for the screen as

"2001: A Space Odyssey," written in collaboration with Stanley Kubrick. Clarke was instrumental in bringing the motif of the alien to the mass market. Clarke expounded on his visions and ideas in more than seventy books of fiction and non-fiction. In 1986 he was named a Grand Master of the Science Fiction Writers of America.

2001: A Space Odyssey might have remained a little-known story lost on the bookshelves, were it not for the adaptation of the book for the screen, and the ability of film to use the themes of space exploration and aliens to captivate the public's imagination. 2001: A Space Odyssey was written when space exploration programs in the U.S. and abroad were nascent projects, percolating in the minds of only a few. In order to expand these programs, broad public support was vital, as the tax dollars of ordinary citizens would fund these projects.

2001: A Space Odyssey won an Oscar for special effects. The realistic picture of space offered by this movie sparked the public's interest and imagination in a remarkable way. Kubrick's direction and screenplay demonstrated the intriguing possibilities of space exploration, vividly showing how lengthy journeys involving suspended animation might actually occur. Clarke's descriptions of the specific maneuvers needed for such exploration are surprisingly accurate. Clarke even envisioned the zero-gravity toilet! The movie itself has been influential for many of today's top thinkers and inventors. Bill Gates, for example, has suggested that the movie helped to form his vision of the computer. It has been said that sections of the film were shown to NASA astronauts in training. With the help of Arthur C. Clarke, public interest in space exploration was stronger than ever before by the 1960s. People were curious, interested, and very much on board.

In the scientific community, however, Clarke is most well known for the invention of the communications satellite. Clarke served as a radar instructor and technician in the Royal Air Force from 1941 to 1946. He was the chairman of the British Interplanetary Society from 1947 to 1950 and took up the position again in 1953. Clarke won numerous awards for his scientific work, including the Franklin Institute Stuart Ballantine Gold Medal in 1963, the UNESCO-Kalinga Prize for the Popularization of Science in 1961, and a nomination for the Nobel Prize in 1994. In the 1950s and 1960s, Arthur C. Clarke was a remarkable ambassador for space exploration. His training in engineering allowed him to envision what space exploration might actually look like, with astounding clarity and accuracy. His background in radar led him to be the first to propose using artificial satellites for global telecommunications, global television, and meteorology.

On April 6, 1965, Intelstat I Early Bird, the first commercial geostationary communication satellite, was launched into orbit. This was twenty years after Clarke had first proposed the idea in Wireless World magazine. When Clarke's prediction was published in 1945, hardly anyone took it seriously. The article articulated the basics behind the launching of artificial satellites in geostationary orbits in order to relay radio signals. Many people thus refer to Arthur C. Clarke as the inventor of the communications satellite. Perhaps to the disbelief of those skeptical readers of Clarke's proposition over sixty years ago, today satellite technology is extremely widespread. At a distance of 36,000 km above the equator, there is an orbit where hundreds of satellites are revolving around the Earth at the same rate that our planet spins. Many refer to this orbit as the Clarke Orbit or the Clarke Belt. To someone watching from Earth, a satellite in geostationary orbit appears to be standing still, but really it is revolving all the way around the planet, once each day. These satellites are relatively inexpensive and extremely useful for myriad communications applications.

President John F. Kennedy

Let us turn to another important link in this chain of human minds: President John F. Kennedy. On May 25, 1961, President Kennedy gave his famous "Special Message to Congress on Urgent National Needs," otherwise known as the "Man on the Moon" address. In this speech, Kennedy asked the U.S. for "531 million dollars in fiscal '62" and "an estimated 7 to 9 billion dollars additional over the next 5 years" to make the dream of landing a man on the moon come true. Some felt this dream was impossible. Others assumed that this move was simply a sly Cold War strategy. The speech followed on the heels of the Bay of Pigs incident, a humiliating event to Kennedy and his administration. As well, the Soviet Union had already made the first strides towards space exploration when they launched the Sputnik satellite into orbit in 1957 and, four years later, Soviet cosmonaut Yuri Gagarin became the first human in space. Three weeks before President Kennedy gave this speech, Alan B. Shepard became the first American in space. Kennedy was feeling an urgency to send a man to the moon.

In Kennedy's own words:

"...if we are to win the battle that is now going on around the world between freedom and tyranny, the dramatic achievements in space which occurred in recent weeks should have made clear to us all, as did the Sputnik in 1957, the impact of this adventure on the minds of men everywhere, who are attempting to make a determination of which road they should take..."

"I believe that this nation should commit itself to achieving the goal, before this decade is out, of landing a man on the moon and returning him safely to Earth."

"No single space project in this period will be more impressive to mankind or more important for the long-range exploration of space; and none will be so difficult or expensive to accomplish."

"...in a very real sense, it will not be one man going to the Moon—if we make this judgment affirmatively, it will be an entire nation. For all of us must work to put him there."

"This decision demands a major national commitment of scientific and technical manpower, material and facilities, and the possibility of their diversion from other important activities where they are already thinly spread."

"It is the most important decision that we must make as a nation."

"...all of you have lived through the last four years and have seen the significance of space and the adventures in space, and no one can predict with certainty what the ultimate meaning will be of mastery of space."

"...while we cannot guarantee that we shall one day be first, we can guarantee that any failure to make this effort will make us last. We take an additional risk by making it in full view of the world..."

Kennedy's vision, outlined in this epic speech, guided NASA's space flight program from the onset. The Mercury, Gemini, and Apollo missions were all designed with Kennedy's goal in mind. On July 20, 1969, NASA fulfilled the mission that Kennedy had laid before the nation. On that day, Apollo 11's lunar module Eagle landed on the moon with Neil Armstrong and Buzz Aldrin on board, and Michael Collins orbiting above. Six hours after landing, Armstrong walked on the moon's surface, speaking the famous line:

"That's one small step for man, one giant leap for mankind."
Aldrin stepped onto the moon shortly after, calling the landscape:
"magnificent desolation." Neil and Buzz planted a United States flag
and left a sign that read, "Here men from the planet Earth first set
foot upon the Moon, July 1969. We came in peace for all mankind."

And lo and behold, in the nick of time for the lunar landing of Apollo 11
in July 1969, a sequence of launches was completed that put satellites in
space over three ocean regions (foreseen by Clarke twenty-five years prior),
which allowed television coverage of Armstrong's extraordinary moonwalk
to be broadcast all over the world.

Transistor Innovation

Bell Labs was the name of AT&T's research lab. In the early 1900s, AT&T
bought a patent that allowed a signal to be amplified regularly along
a telephone line. This meant that a telephone conversation could go
on across any distance, assuming there were amplifiers along the line.
However, the vacuum tubes that made this possible were not very reliable,
sucked too much power, and produced an excess amount of heat. In the
1930s, Mervin Kelly, the director of research at Bell Labs, realized that an
improved device would be necessary for the telephone business to keep
growing. He thought that the necessary improvement might be found in a
type of material called semiconductors.

A semiconductor is a group of materials whose conductivity is in-between
metals and insulators. (Not, for example, a metal such as copper that
definitely conducts electricity or a material such as rubber, which definitely
does not.) The conductivity of semiconductors can be changed by injecting
charges into these materials or shining light on them. After many years of
research involving many brilliant minds (and much internal controversy),
on June 30, 1948, Bell Labs unveiled the solution and a replacement for the
vacuum tube. They called this invention the transistor. It indeed allowed
an electrical signal to be amplified, and answered the problem of the
unreliable vacuum tubes. It replaced the vacuum tubes in telephone lines
and eventually became the most important component of the electronic
age. At the time, however, no one paid much attention to it, and it was not
found to be of use in any consumer products, save the hearing aid.

Unsatisfied Wants of the People

In 1952, Akio Morita of Japan purchased a license from AT&T to begin building transistors. Engineers at Bell Labs informed him that the transistors were only good for making hearing aids. However, Morita wanted to use the technology to make small radios.

"But radios are far too expensive to devote to just one person," Morita's engineers told him.

"Then we will manufacture transistor radios so they are not too expensive for just one person to use at a time," Morita replied.

"But there are not enough radio stations to support such an idea," said the skeptics.

Morita answered: "There will be."

In the 1950s and 1960s, Masuru Ibuka and Akio Morita (who had recently founded Sony Electronics) began mass-producing small transistorized radios. All of a sudden, the transistorized radio changed the world. For the first time, information could spread quickly and easily to almost every remote corner of the Earth. Over the years, Morita's company created a long list of new technology: the AM transistor radio (1955), the pocket-sized transistor radio (1957), the two-band transistor radio (1957), the FM transistor radio (1958), the all-transistorized television set (1959), the all-transistorized video tape recorder (1960), and the small screen transistorized television set (1961).

People at Bell Laboratories, along with all of the electronic manufacturers in America, had decided on some level that the consumer was not ready for transistorized equipment, or perhaps that the equipment was not ready for the consumer. Sony thought outside of the box in this instance and asked: "What are the unsatisfied wants of the consumer?"

Sony took note that young people were lugging heavy phonographs and battery-powered radios with audio tubes on picnics and to the beach. Listening to the radio in the 1950s was only easy to do from home. Sony identified a new growth market, satisfying the unmet needs of the consumer with portable transistor radios, and within a remarkably short period became the worldwide leader in this market.

The broadcasting stations of this era also lacked vision. Radio stations just assumed that people would need to be in their homes to listen to the radio, and thus the number of hours in the day when people would actually be listening was quite limited. Consumers didn't demand a mobile radio, and no American companies identified the need, except Sony. In order to meet this goal, they began looking for the appropriate technology. Once they identified the technology, it was only a matter of time before they launched their highly popular mobile radio.

Today we find ourselves in a similar place within the field of mobile Internet technology. The technology exists to provide widely accessible mobile Internet to our communities, and yet we have not put this technology to use for ourselves on a major scale. Let us be inspired by these visionaries of the past, and by these remarkable accomplishments that I've just written about. In this book I hope to create a collective vision for a mobile Internet technology that will allow all of us to connect to the Internet easily and quickly from just about anywhere -- a remarkable accomplishment of the future.

OVERVIEW

"A new generation of innovation is about to change the way technology interacts with people … In the next few years we are going to take a leap into uncharted territory." **Steve Ballmer, Microsoft CEO (February 17, 2009).**

In this chapter I will expound upon the necessary components involved in creating an Internet infrastructure that can be accessed from just about anywhere. In this book, I will refer to this system as Web on-the-go. Our society indeed already possesses all the technology required in order to implement this Web on-the-go infrastructure. I will explain all of these contributing elements, to the best of current knowledge, in the following pages.

Currently, we have a wide range of options for connecting to the Internet from our homes. In order to access the World Wide Web from home, we need only two things: a computer, and some sort of Internet service, both of which are easily accessible, given that one is able to afford them. Computer systems include IBM PC, Macintosh, Dell, desktop, and laptop. There are a variety of options for software platforms, including Windows, Mac, and Linux.

Similarly, we have quite a few options when choosing a home based Internet service. We can purchase DSL through the phone lines from AT&T, or from cable providers such as Comcast,

or even satellite providers. From home, all of the choices are ours, and we have plenty of options. It is important to note that in this Web on-the-go model, we, as individuals, in some ways, do not have as much choice, nor do we get to decide everything. Access to the Web could become much more universal, and not saddled with so many personal decisions. Because this technology is applied to entire cities or counties, there is a wide array of people who must be involved in helping to make this vision a reality, such as city council members and city executives.

So what systems and technology are necessary for our communities, in order to access the Internet from anywhere, to connect to the Web, on-the-go? Again, we need two things: a computer, but smaller this time, mobile, and hand-held. And, we need an Internet service, wireless, strong, and available everywhere. So you can see, it is not necessarily that the players in the game change, but that we implement dramatic and far-reaching innovations, and in the process change the entire playing field. In my mind, I see that there are four major components to this project:
- Mobile device component
- Software platform and applications component
- Wireless infrastructure component
- People component

Mobile Internet Devices (MIDs) are a relatively new advancement in Internet technology. These devices are small and mobile, and allow a user to communicate with others, view movies, and access information from the Internet. Electronics companies are employing new Intel CPU technology (more on this later) to create pocket-sized products that allow people to easily access the Web-on-the-go. These Mobile Internet Devices have long battery life, and continue to get smaller, more stylish, and lighter as we move into the future.

These devices allow the user to surf the full Internet, keep in touch on instant messenger (IM) or voice over IP (VOIP), watch high-definition (HD) video and audio, connect with builtin wireless around town and around the world, experience broadband on-the-go for wireless access beyond hot spots with select WiMAX-ready MIDs, and get detailed directions and personalized information, based on a specific location. The possibilities for this technology are endless, and these devices are a pivotal component to a complete Web-on-the-go Internet system.

As with any new technology, Mobile Internet Devices provide many opportunities for new growth. For example, Moblin, an open- source Linux project, allows users to share software technologies, projects, code, and applications specific to MIDs. In the past few years, mobile device technology has accelerated rapidly. Not long ago, cell phones had only two components: a mouthpiece and a speaker. Today, however, mobile devices have evolved into powerful handheld computers that contain cameras,

GPS, Wi-Fi, Bluetooth, keyboards, and even elegant screens with touchpad features. Microsoft's CEO Steve Ballmer has predicted that in a short time these devices will be so lightweight and thin, they can be rolled up and slipped into a purse or briefcase.Examples of this new mobile technology (sorry, no roll-up mobile devices yet) include Windows Mobile Device, Apple iPhone, Google Android, Blackberry, Sony Pocket PC, NetTop, and Samsung MID (Mobile Internet Devices). These devices are helping change the face of Internet access. But how can something so small be so powerful?

Inside each of these mobile devices is a little gem called a CPU, or central processing unit; a CPU is the electronic circuit in a computer that can execute the computer's programs; every computer has one. The CPU in a mobile device is just really, really small. The Intel Atom processor is the smallest processor built by Intel. It is built with the world's tiniest transistors and was created specifically for simple and affordable mobile devices, netbooks (a class of laptop computer designed for wireless communication and access to the Internet), and nettops (a type of mini desktop or small-form factor computer). These devices can be used for photo and video viewing, e-mail, messaging, browsing, social networking, VOIP, and many other Internet applications.

Another interesting component of the many mobile devices is what we call solid-state drives (SSDs). SSDs can be compared to traditional magnetic media drives (commonly known as hard disk drives or HDDs). However, SSD technology usually allows a system to perform much faster than a traditional hard disk drive. There are no moving parts in SSDs, so they are at hardly any risk of mechanical failure. SSDs also consume a great deal less power than a traditional hard disk drive while simultaneously offering "improved overall system responsiveness." The SSD technology is "cooler and quieter" than that of HDDs. This SSD technology is currently offered by Intel for use in netbooks, nettops, mobile internet devices, and digital entertainment.

Microsoft, Apple, and Google have created software platforms for these new mobile devices. This platform provides tools for software developers to build any applications that could be installed on mobile devices. Thus, software developers all around the world have been building applications for the devices that are downloadable from the Web. Google has provided a marketplace called Play Store for these software developers to sell their mobile device products. Some of the software is even offered free.

There are three salient types of applications currently available for mobile devices: standalone applications, Web applications, and complex applications such as Google Latitude. I will briefly discuss each application.

Standalone Applications

You download these applications onto a mobile device. Standalone applications can be run without any networking connection. Simple examples of these applications are games such as chess and checkers. This type of software does not need an Internet connection to function.

Web Applications

This type of software for mobile devices does need an Internet connection to function. Good examples of these applications would be using e-mail, watching YouTube, and accessing Web browsers. You can launch these applications from your mobile devices when there is an Internet connection. If you are at home, your mobile device uses your home Internet connection for these applications. At home, you may have an AT&T Internet connection with a wireless router. Your mobile devices can connect to these wireless access points and make the applications on the mobile device function. If you are at work, then your mobile devices use your workplace network connections. If you visit Starbucks, then these mobile devices use the Starbucks wireless hotspot connections. If you are walking on Castro Street in Mountain View, the city provides Wi-Fi access points on the street lamp posts. Your mobile devices can connect to those city Wi-Fi access points and you are good to go with your Web applications. If you are on public transportation, then you can use the wireless connection provided to access the Web. Ultimately, a mobile device has freedom, and one can use it like a laptop computer. Right now, I cannot use my cell phone if I leave the United States. But with a mobile device, I can use it anywhere under the sky on planet Earth.

Complex Applications

Recently, T-Mobile launched the Google Android Mobile Phone. This device has a Google Maps application. One can launch this application and then select My Location. The device has a GPS receiver that helps to identify the coordinates of one's current location; that information is passed over to the Google Maps server; immediately, the map is displayed on the mobile device. This is an example of a complex application. The mobile device uses GPS to locate the coordinates and access the Web for map details.

Android has another advanced application called Latitude; anyone can sign up with friends and family. Assume that the children in a family have these mobile devices. This Latitude application periodically (say, every 30 minutes) sends the location information of the device to the Google server; a mother can access the Web and figure out the current location of her kids, whether they are still in school or walking back home or on the playground; it has a privacy control option as well.

Right now, with applications like Chat, we know exactly who is available online. Now, with Latitude, you can see the current locations of your friends. One can envision a conversation such as, George says, "Hi, Jerry, are you in San Francisco? I am in Oakland. Can we have lunch together in the city?" Jerry can wait in the restaurant in San Francisco and watch George coming closer to the restaurant while he offers precise directions to the location.

This application also comes in handy in a situation I'm sure we are all familiar with: finding our lost car in a big parking lot or city. For example, let's say you go to a concert at the Shoreline Amphitheater; you parked your car in the open parking lot; there are no identification marks in the lot. When the concert is over, you need to reach your car but you can't locate it. Assume that your car also has a mobile device installed. This Latitude application can capture your car's location. On your mobile device, you can see the car's location and then walk to it.

To make this happen, car manufacturers, software developers, and business owners need to know about this particular unsatisfied need of the customer; car manufacturers need to know about the power and features of these mobile devices; software developers need to know about specific applications such as Google Latitude and capturing the location of the car to take the customer closer to his car. Finally, the Shoreline Amphitheater authorities need to be in the know so that they can provide the Wi-Fi infrastructure in those parking lots.

Networking Infrastructure

One can also use these mobile devices to communicate with equipment at the gas station. Instead of sliding a credit card, one can pass information to the credit card equipment through wireless communication such as Bluetooth. We will see the effectiveness of this system in much more detail in the coming chapters. Wireless technology is by no means a brand-new phenomenon; to exemplify how long wireless has been playing a pivotal role in our lives, let's take a look at a device that no home has gone without for over half a century, the radio. From its inception, the radio was using wireless as its sole medium for transmission; from its first moment, the radio broadcasting industry was always "on the air." A few decades ago our grandparents were listening to radios that had a big vacuum tube receiver and a big antenna. Because radio broadcasting was completely wireless and available pretty much everywhere, even on the road, it was incredibly easy for Sony to launch the pocket transistor radio that I spoke of earlier; all they needed to invent was the small device that would receive the broadcast. In other words, they did not have to worry about creating an infrastructure to support their invention; it already existed! Television broadcasting began in the same way, completely wireless; all one needed to watch TV was a television and an antenna. However, at some point, private cable television was introduced for interested subscribers, which added a "wire" to the equation.

Perhaps the telephone could have followed a similar pattern, using wireless media instead of conventional bell wires from the beginning. Following this train of thought, had there been no wires for communicating via telephone, then we never would have used phone lines to connect to the Internet (as many of us still do with dial-up and DSL). And the same goes for cable television. Had there never been private cable television broadcasting, the Web would not have been transmitted via Comcast cable. If this had been true, then the Internet and the Web would have started as wireless media and would have been much easier to access all along; right from the outset, we might have had an antenna at home to receive an Internet connection. At this point in history, we are realizing the benefits of wireless Internet, and now we are exploring ways of converting the old infrastructure from wired to wireless; right now, we use extensive wires, cables, and connectors in order to access the Web. Those wires have copper or aluminum metal inside and are coated with plastics; there are huge factories manufacturing these wires, big trucks transporting the wires and cables; laying them over the poles or underground is always disruptive and challenging. Wireless is a "green" solution.

Next I will discuss the key elements involved in a Web, on-the-go wireless infrastructure; I want to point out some current limitations that exist within this system; I've found often we don't even question these limitations; the system is all we've ever known. Let's take a look at where we've been, where we are now, and where the vision lies for the future.

Cellular Phone Network

In 1947, Bell Labs engineers at AT&T invented cells for mobile phone base stations, and Bell Labs continued to develop this technology during the 1960s. In 1945 the zero generation (0G) of mobile phones were introduced. These phones were not officially known as mobile phones because they didn't allow a user to move from one cell (coverage area) to another without interruption. It wasn't until 1984, when Bell Labs invented what is called a "call handoff" feature, that mobile phone users could travel in between different cells in the course of one conversation. Motorola is widely recognized as the first company to introduce a practical mobile phone that could be used outside of a vehicle setting. Motorola manager Martin Cooper dialed the first call on a handheld mobile phone on April 3, 1973. Since then, due to the low costs of establishment and quick deployment, mobile phone networks have spread like wildfire around the globe. The first such commercial cellular network was built in Japan in 1979 by NTT. In the early 1980s, fully automatic cellular networks were introduced, commonly referred to as 1G (first generation). This led to a boom in cell phone use, especially in Northern Europe. This was followed by the digital 2G (second generation), launched by Radiolinja (currently part of Elisa Group) in 1991 in Finland. Today, both mobile devices and the infrastructure to support them are experiencing an explosion in growth all over the world.

However, the current cellular network is an aging technology, especially when compared to the Web and the Internet; most people don't realize this, and in recent years some have been trying to push the concept of accessing the Web via the cell phone; much time and resources have been wasted trying to make the Web work on the existing cellular network. Companies are attempting to push powerful Web programs like Google Maps on these very small mobile devices, using the antiquated cellular network. This has been a very expensive undertaking, and most of the time, the outcome has not been desirable. The current technology is referred to as 3G or 4G, and it is outdated. The cellular companies came up with this system; it is very expensive to install and maintain, and it is undoubtedly an outmoded manner of thinking about cellular infrastructure; with a system already in place for cellular voice communication, companies are attempting to use this same technology in order to provide Web access through the cellular network. Once again, if we play the "what if " game and imagine that the current cell phone network had never been created, then as early as the year 2000, people would have begun to explore Wi-Fi and other wireless technology for the mobile Web.

Cell phones have indeed become an irreversible part of our daily lives; and yet cellular service as it exists today is not only outdated, but comes tethered to many restrictions; what if many of these restrictions ceased to exist? Think about it. As the system is now, one must purchase the device (cell phone) only from the actual service providers; one must sign a two-year contract; one must check the time of day, and the day (weekday vs. weekend) before feeling free to make (or receive) a call. Most plans offer a limited number of minutes; there are penalties if ever a subscriber exceeds their allotted minutes or breaks the two-year contract; there are extra fees for sending and receiving text messages or accessing the Web. This system offers the consumer no freedom at all.

A story that illustrates the limitations of the current cell phone network involved a man named Wayne Burdick, on a cruise ship in the Miami Harbor; while the ship was still docked in Miami, Burdick set up his laptop and wireless card and accessed his Slingbox device; this enabled him to watch a Chicago Bears game via the Internet; he was watching the game for about three hours. Burdick returned home to find a bill from AT&T charging him over $27,000 for the three hours of Internet usage.

Stories such as this, though funny, feel all too close to home; who hasn't paid an exorbitant cell phone bill? I would propose that this sort of limitation on our cell phone usage is not necessary.

However, in order to get there, we must evolve toward an up-to-date wireless infrastructure. The solution lies in new technology and new paradigms of thought, both of which I will continue to expand upon throughout this book.

Wireless Hotspot/Municipal Wi-Fi

One example of newer wireless technology is a Hotspot; this term refers to a business or other venue that offers Internet access over a wireless local area network (LAN). Wi-Fi hotspots were first proposed by Brett Stewart in San Francisco in 1993; he didn't use the term "hotspot" at that time, but called his idea "publicly accessible wireless LANs." The original idea for a wireless Hotspot was that users would pay for broadband access at these specific locations. These days, just as often, a wireless Hotspot is a free service; Hotspots continue to grow around the nation and the world. Using the same technology, wireless networks that range across entire cities, often referred to as Municipal Wi-Fi, have been gaining popularity. MuniWireless reports that over three hundred metropolitan wireless projects have been started.

Wireless on Public Transportation

Accessing the Internet on public transportation is already happening in many places around the globe. In Japan, citizens can log-in on many public trains and buses. In the U.S, there are at least twenty different cities offering wireless Internet on public transport, including Colorado Springs and Cincinnati. Buses running from Heathrow Airport in England are also offering free wireless Internet. "Given the passenger demographic, airport bus and coach routes are a natural choice for Wi-Fi services," said Dave Palmer, senior vice president of worldwide sales at Icomera (a company specializing in mobile Internet technology). "Business passengers can get online at no cost and make the best use of their time. More importantly, free Wi-Fi is the preferred way to connect while traveling abroad, as it avoids high international 3G data roaming charges."

WiMAX

Another wireless option is called WiMAX, is a broadband wireless system that is widely supported by both the computer and telecom industries around the world, which has helped to make this technology particularly cost-effective. It has been created in order to offer significant business benefits to operators and users in a wide array of environments (enterprise, consumer, emerging, public service), geographies, and demography (urban, suburban, rural), both over the short- and long-term.

Wireless over Unlicensed Spectrum

There is a phenomenon called white spaces; these are slivers of wireless spectrum between the broadcast channels used by television stations; originally, these slivers were created in order to avoid interference between television broadcasts. However, on February 17, 2009, television stations moved on to new frequencies, following an order by the government to switch everything over to digital broadcasting. In this change, many are glimpsing an opportunity.

A few companies, including Google and Microsoft, have been in conversation with the Federal Communications Commission, with the hopes of offering up this spectrum to the public for free, unlicensed use. Other companies are also in favor of freeing up these airwaves, including Dell Inc., Intel Corp., Hewlett-Packard Co., and the North American unit of Philips Electronics. Google has been clear about the benefits such use of the white space would have upon their own company; wherever the public is gaining access to the Internet, an ad-based company such as Google stands to benefit. Silicon Valley has said they would also benefit from such a system, and no doubt the other companies on board see it as a lucrative endeavor for them as well. There are already similar slices of airwaves being used by home, business, and city Wi-Fi networks. An executive from Google has referred to this plan as providing something similar to "Wi-Fi on steroids." Google has said publicly that the airways could provide "huge economic and social gains if used more efficiently."

Google believes that mobile devices using this white-space spectrum should be available by the end of 2009; Google sees this white-space spectrum as a fantastic air space in which to operate a brand new type of cell phone and wireless device based on Android, a software model that many other companies are now planning to replicate; using the proper mobile device and the white space, it is thought that users will be able to watch movies and do other things while on-the-go that are now difficult on slower networks.

This idea is still opposed by broadcasters and manufacturers of wireless microphones. These people still fear that the mobile devices using the white space would cause interference. The FCC is currently conducting tests to gain a better understanding of whether this white space spectrum can be used without disrupting television broadcasts. Google has offered a few solutions to the argument that the use of the white space would cause interference; they proposed keeping a few channels for the exclusive use of wireless microphones, along with medical telemetry and radio astronomy devices; they also offered that "spectrum-sensing technology," already being used by the U.S. military, can sense whether a channel is being used before accessing it, which would avoid interference between channels. Google has also offered to help other companies take advantage of the white-space airwaves.

The FCC is currently studying a proposal that would create two categories of user groups for these airwaves: one for personal portable devices that are low-powered, and another for commercial operations.

Satellite Wireless

The VSAT (very small aperture terminal) was invented by Hughes in 1985. Soon after, the company introduced to the market the satellite networks industry, with Wal-Mart signing up as their first customer. The company is a world leader in mobile satellite system design and development. Over the years, the company has continued to expand, and currently Hughes ships satellite terminals to a variety of customers in over one hundred countries. Hughes technology is available in North America directly from Hughes. In other countries, their satellite networks are available from various resellers. The company's broadband satellite networks and terminals are based on the IPoS (IP over satellite) global standard, approved by the TIA, ETSI, and ITU standards organizations.

Ad-hoc Network

Ad-hoc technology was originally developed in the military; it became necessary to create a system for troops to communicate in regions where there is no cell phone infrastructure. Imagine, for example, that hundreds of U.S. troops are moving in different locations around the deserts of Iraq; it is imperative that they be able to communicate among themselves to effectively implement strategies and plans; their location is both foreign and remote, and lacks any Internet infrastructure. In order to resolve this problem, the military created what is called the ad-hoc network; this technology is very simple; each unit, or military division, has a wireless device; each device is able to make a wireless connection with neighboring devices, and thus a network is rapidly established.

I think that this ad-hoc network technology will prove extremely useful in domestic situations. For example, let's assume there is a conference underway for software developers; even within the conference hall, an ad-hoc network can be established; this will allow all the participants to be connected, without relying on the Internet. Or let's imagine that a family takes an outing to a large store such as Costco; while the husband peruses the computer section, the kids are checking out the video game section and the wife is in the grocery section; they are all under one roof and yet if they want to communicate, they need to pick up their cell phone and connect via their cellular network; this ad-hoc network would allow the family to be connected easily. "Yes, I am done. Come meet me at the check-out line," the wife could broadcast quickly and efficiently to her family.

As you can see, the technology exists to upgrade our aging wireless infrastructure, and to provide an easily accessible Web on-the-go system that will undoubtedly prove to be an invaluable service to communities everywhere.

Social Responsibility

It is interesting to me that Arthur C. Clarke had one vision that proved extremely useful to the general public (satellite communication) and another vision that might not necessarily be seen as "useful" in that way (man on the moon). Though arguably, putting a man on the moon has served humanity, as it filled us with wonder, optimism, and inspiration. Visionaries should not spend all of their creative energy envisioning things that are not helpful to humankind. One ought to dream with social responsibility, if you will. However, it is not useful for us to tell these dreamers what to dream about. Once a condition is placed upon them, a protest ensues, and the dreamer might spend an entire lifetime fighting the limitation that has been placed upon them. In the case of Arthur C. Clarke, one might try to place a higher value on one vision than on another. Perhaps one might think that the communications satellite is more worthwhile in the long-run. As I see it, both of these visions came from a common action; Arthur C. Clarke was incessantly looking upwards at the sky and dreaming of venturing into space; perhaps his vision of a man on the moon led him to explore the possibility of satellite communications. What if Arthur C. Clarke had spent all of his time looking down at the earth? Maybe he would have dreamt instead of laying cables in the ground for distance communication.

We live in challenging times, with many changes looming in front of us. It is not the time to be thinking frivolously, or only about ourselves. In my eyes, dreaming with social responsibility is extremely important, and I would offer it as a suggestion (not a condition) to all the visionaries of our time.

Common Mission

Usually, a visionary creates a document of their vision. This could be in the form of an article, a paper, or a book. In preparing this documentation, it would be most efficient if they incorporated the perspectives of policy makers, scientists, management, and the general public. The document should communicate very clear messages to each sector listed above. The different people involved in Web on-the-go will need to work together just as members of a football team. Everyone has his or her specific role and responsibility, and everyone shares a common goal. Provided that the importance of each team member is fully recognized, and the overarching objectives are clear, then everyone will be able to contribute effectively to meet the common goal. As in most areas of life, effective communication will be critical.

Oftentimes, visionaries live in a world that is isolated from the policy makers, while scientists live in a very separate reality from management executives and others. We all find ourselves in a bubble reality, paying little attention to what is happening in other sectors of society, and often criticizing another sector for not doing their job well. But for a project to be successful, each group must recognize the importance of every other group, and learn how to communicate effectively. In the business world, mission statements act as a guiding force for all the people in the organization. The mission statement helps visionaries to dream about a particular field. For example, Mobile Web helps the policy makers, management executives, scientists, and engineers in coming up with their own unique objectives. Ideally, the general public also becomes informed about the final product and services that each individual will ultimately receive from this mission. Once the objectives are clearly defined, then everyone will, hopefully, come up with the resources and time needed to achieve their common goal.

So the document created by the visionary must communicate clearly with the general public, the policy makers, scientists, and others. Included in the document must be details needed for each sector to move forward effectively. The politician must be shown how to make a new political decision and how the decision will be seen as a historical one. Similarly, the scientist must be told about the more intricate details of the project (such as with the zero gravity toilet, for space travel). As well, the management executives must be able to clearly envision the resources needed to complete the project and how to market these products to the public. The document should not, however, go into the details of implementation. This is the job of the management executives and others.

The next few pages briefly touch on some of the key players who will need to be involved in Web on-the-go technology.

Visionaries

Visionaries don't think, they dream. They don't live in the present but instead seem to reside permanently in the future. It could be said that they don't follow logical thought patterns, and yet their imagination is always at work. Often it seems that visionaries don't care about financial, human, and other resources, or about the feasibility and practicality of their ideas. It appears that their imagination has no boundaries or constraints. Once an idea takes shape in the imagination of a visionary, it is strongly believed to be real. Oftentimes visionaries put themselves at the center of their imagination. For a visionary, once one has documented their vision, there is a feeling of completion. There is a sense of satisfaction and achievement. It really doesn't matter whether someone takes their vision and makes it into a real object or experience. For example, Clarke's vision was "Arthur C. Clarke is on the Moon." Once this vision was complete, then it became documented with more particulars. Since Clarke imagined himself as the subject of his vision, he captured all the experiences of space travel with astounding accuracy; an example was his zero-gravity toilet idea. Since the visionary resides in the future, it is often difficult for others to realize, understand, and appreciate his ideas. More often than not, new ideas are scorned and categorized by the logical masses as hypothetical and far-fetched. It may take decades for other people to fully grasp the ideas of forward-thinking visionaries and to set about making these ideas manifest.

To speak broadly, quite often people are totally consumed with their own lives and problems, and do not have space in their agenda to ponder new ideas. For example, right now we find ourselves in the midst of an economic recession. Many are spending a great deal of time analyzing the past and trying to understand what went wrong, and when. Others are spending their time coming up with plans and methods for whatever challenges lay ahead. We are so wrapped up in the past and the problems that we face now that we have a difficult time seeing more than ten or twenty years down the road. Visionaries don't seem to be daunted by the past or by present crises, or by challenges to come. All they see are future opportunities.

At this moment they might be struggling to pay the bills and buy groceries, and yet they are still dreaming about putting a man on the moon, for example. First and foremost, these creative and innovative people must identify their innate gift of vision, and keep these imaginative processes flowing. More often than not, visionaries don't really understand their gifts. They often beat themselves up, thinking of their dreaming tendencies as a handicap, rather than having a positive impact on the world. Indeed, our society mirrors this attitude, and perpetuates it.

And others as well must begin to recognize the importance of these visionaries in our society, and create a space for them to continue with their dreams. Business establishments need to identify these individuals and set them free, providing them with the resources necessary to continue to dream. For instance, the most salient visions of Arthur C. Clarke was putting a man on the moon and launching a satellite for communication. It is possible to appreciate the importance of these ideas only when we think about what life might be like if these visions had never come to fruition. Had Clarke never imagined a man on the moon, Russia and America most likely would not have attempted to build rockets or send human beings into space. NASA probably would not exist, with its world-shaking achievements. Visions such as these have been the most important motivating factor for countless innovations and breakthroughs in our history.

Policy Makers

Policy makers are politicians. The main objective of a politician is to secure a favorable place in history. Kennedy, of course, will always be remembered as the president who sent a man to the moon. Policy makers need innovative ideas that will have a huge impact on society. Policy makers are (or should be) always on the lookout for the visionaries of their day. Politicians have resources, both financial and human. When they choose the right vision, and effectively implement it with the resources available to them, then they will secure their place in history. Kennedy chose the man on the moon vision. Politically, Russia's prior achievements in space exploration led the president to make the decision to land a man on the moon. He made the best use of the resources available to him at that time. And, when Armstrong indeed set foot on the lunar surface, every single citizen felt a deep sense of achievement and pride.

In the face of the 2008 economic recession, President Obama has chosen to focus toward a vision of broadband access for all. Using resources from the stimulus package, Obama hopes to make this vision a reality. Once this dream is realized, it will have a tremendous impact on the lives of Americans, and people all over the world. In the United States, the president makes policy decisions at a very high level. In the stimulus package, broadband for all means for all, every city and town. However, it is up to city councils and city government executives to decide how to implement such a policy.

It is imperative, then, that these more local policy makers have substantial vision and knowledge about the mobile Web topic in order to understand all applicable opportunities, risks, and challenges.

Let's take the case of municipal Wi-Fi for the city of San Francisco. This idea was proposed initially by executives of Google and Earthlink. City executives asked for clarification on certain topics, and when Google and Earthlink failed to respond, the project was dropped. In this case, executives from the technical firms did not respond to the city executives' queries, and thus the entire project was lost. I think that those technical firms must realize that they missed out on a fantastic opportunity. Had they communicated their ideas more clearly and diligently, I think it's safe to say that the mobile Web would be up and running in 2009 in San Francisco.

Without the support of the policy makers, the mobile Web cannot be implemented. Let's not ignore them! Let's work to answer all of their questions and be sure to clarify for them all of the opportunities, challenges, and risks involved in this project. Let's treat them as partners.

Scientists and Engineers

The Bell Labs story might offer an important lesson to current scientists and engineers. Oftentimes, engineers and scientists isolate themselves from the real world and focus exclusively on their own inventions. They get so caught up in what they are working on that sometimes they can't see the bigger picture: mainly, how will their invention serve society, and what are some of the practical applications?

The engineers at Bell Labs were not very receptive to the ideas of Sony's Akio Morita, which they thought were not practical. Had they given him the benefit of the doubt, however, and worked with him to achieve a common vision, perhaps the end result would have been even more valuable, more lucrative, and more helpful to society than the transistor radio.

If management executives challenge scientists and engineers to come up with a new product, then those scientists will work diligently to invent that new product or solution. Once a product is invented, however, and a prototype is made, many scientists think that their job is over. Often they are not so concerned with how the product is manufactured, marketed, or used. Or they are somewhat removed from the problem or challenge that the product they are creating is attempting to solve. Because of this, it's tricky for scientists and engineers to fully understand both the satisfied and unsatisfied wants of the consumer. Ideally, scientists and engineers would value the importance of the other key players in the marketplace, and work to communicate with them throughout the process. Once this basic understanding is established and communication is made effective, then it will be much easier to make a vision into a reality. It will be a win-win for all members of the team.

Let's return to an earlier example: Google Latitude. Scientists know how to capture GPS location data; they know how to update current location information; they know how to provide the current location of friends and family members. But it was not a scientist who came up with the idea to build a Google Latitude application. Scientists did indeed play a key role in the development of this software, but someone else, who understands the value of this application, clearly defined the requirements. Once they got the idea, the scientists and engineers used their vast expertise to build that application. It is essential for each member of the team to understand the general qualifications of other team members. Then they can effectively communicate.

Device Manufacturers

Intel has realized the potential for the Mobile Web and is actively promoting processors for Mobile Internet Devices. Many manufacturers like Sony, Samsung, HP, and HTC are already making Mobile Internet Devices, Pocket PCs, and other devices. Nowadays, many people have a desktop as well as a laptop. Likewise, one person may have two or three mobile devices, a Pocket PC in his shirt pocket and a Mobile Internet Device in his briefcase. He may use the small device to read emails and look at maps, and the slightly bigger one for writing emails or typing a document while sitting on the park bench. Soon there will be a huge volume of these devices flooding into the global market. And they all have wonderful features such as a camera, GPS, and other networking connectivity.

Assume that you built a highly sophisticated sports car, but there are no roads or highways or tracks. Similarly, Mobile Internet Devices will be of no use if there are not enough service providers for wireless infrastructure.

Wireless Service Providers

Private establishments such as Starbucks, McDonalds, bookshops, and so many others provide wireless access to their customers. Public transportation also offers wireless access. But city council and city government executives need to make a major decision about providing wireless access all over their city. First, they need to understand the importance and necessity of providing these services to people. Second, they need to be aware of the various technologies and technical solutions available in order to make the right decision.

Again, effective communication is essential among all these team members in order to make the Web on-the-go mission a reality.

Software Developers

Software developers can build pretty much any application. But we can't assume that they will identify all of the potential applications and then deliver this to the public. I have seen software developers come up with newer, better versions of old software (such as chat and email), but it might take a new vision to get them to switch gears.

Management Executives

The work of scientists quite often is contained within the lab. It remains to management executives to take products out of the lab and deliver them to the public as a new product or service. It is absolutely pivotal that scientists understand the importance of management tasks and vice versa. With mutual respect, these different players can grow together. Again, Bell Labs and Sony are examples.

Other Business Establishments

When the Web was launched, it was mostly used for email communication. Only later did business establishments start paying attention and begin to use the new Web technology efficiently for their businesses. The banking industry came up with online banking solutions; the retail industry came up with online shopping; fast food chain restaurants came up with many online features. It is essential for those in industry to take a look at the most current technology and figure out the best possible way to make use of it in order to serve their clientele.

Right now, most of the Mobile Internet Devices are being used to read and send email on-the-go. Very soon, the banking industry and retail and other business establishments will take a look at the ease of doing business available with those Mobile Internet Devices and may come up with some surprising and innovative applications.

General Public

What the public has is purchasing power. If an effective product and solution is offered, most likely the public will recognize its value, and spend money on that product.

Blind Men's Perspective

You have seen now how many different components are involved in this mammoth industry. Each company has a different expertise: Intel has an expert perspective on the chip and SSD. Samsung and HTC have a very informed perspective on manufacturing Mobile Internet Devices. Google, Microsoft and Apple have the best software and applications info around. Similarly, Wi-Fi service providers could tell you all about the infrastructure of such a system.

Here is the blind men and elephant story: Six blind men were asked to determine what an elephant looked like by feeling different parts of the elephant's body. The blind man who feels a leg says the elephant is like a pillar; the one who feels the tail says the elephant is like a rope; the one who feels the trunk says the elephant is like a tree branch; the one who feels the ear says the elephant is like a hand fan; the one who feels the belly says the elephant is like a wall; and the one who feels the tusk says the elephant is like a solid pipe.

A wise person explains to them, "All of you are right. The reason each one of you is telling a different picture is because each one of you touched a different part of the elephant. So, actually the elephant has all the features you mentioned."

This is used to illustrate the principle of living in harmony with people who have different belief systems from yours, and that truth can be stated in different ways. It is known as the theory of Manifold Predictions.

My intent in this book is to eliminate the "blind person's perspective," to bridge the gap between technology and the customer on-the-go, and to focus on the whole perspective. I want to show how all of these groups are holding unique positions in a very large and connected playing field.

The following pages illustrate the potential real-world applications of Web on-the-go through a series of stories. Each story describes a specific life situation in which the characters do not have the Web on-the-go technology. I will then introduce the Web on-the-go technology and retell key parts of the story, exemplifying the ways in which this system would alter, simplify, and benefit the lives of ordinary citizens. Please be aware that the Web on-the-go scenarios are hypothetical, and some of the applications do not yet exist. However, I include only applications that could easily be created, given currently available technical expertise, models, and materials. Indeed my hope, and an overarching goal of this book, is to introduce these ideas to the proper audience, who will then be inspired to make them manifest.

ZOO

As the plane touched down in San Francisco, Julia could not contain her excitement. Descending over the Bay Area was exhilarating, and Julia was astounded at the beauty of the hills, the Pacific Ocean, and the skyline of the city. It was her first trip to San Francisco, and she was here on vacation visiting Jerry, whom she'd met over five years ago when he was an exchange student at her university in London. Jerry and Julia hadn't seen each other for over a year. Julia was very excited to reach San Francisco and immediately wanted to share her excitement with her family back home in London. She had promised her parents that she would be in touch when she reached the United States. She also wanted to let Jerry know that she had arrived safely and on time. Her cell phone worked only in the UK, not here in the U.S. So she couldn't contact Jerry or her family, and they couldn't contact her. Indeed, she realized it was silly to have brought her cell phone at all.

Once inside the airport and through customs, Julia located a pay phone. Fortunately, she had exchanged some Euros into US dollars back in London, but had to go into a restaurant to change a dollar into quarters. She called Jerry, who was just pulling up outside the baggage claim. Elated to see each other, they told stories over dinner and drinks at a Mexican restaurant in Palo Alto, where Jerry lived.

Unfortunately, Jerry had to work the day following Julia's arrival. He was fairly new at his job with a technology firm in Palo Alto and hadn't been able to get the day off. Over dinner, Julia decided that she would go to the San Francisco Zoo by herself the following day. She would take public transportation because she didn't want to take Jerry's car, or rent a car, as she was nervous about driving on the other side of the road, especially in the middle of the city.

Jerry didn't have a clue how to give her advice on public transit, as he always drove his car, though he knew it was possible to get to the zoo by taking buses and trains. After dinner, back at Jerry's apartment, they looked up the San Francisco Zoo from Jerry's computer in order to plan Julia's trip. They were happy to discover that the SF Zoo Website encouraged people to take public transportation, and even offered a discount to those customers who brought their receipt. Jerry and Julia logged on to the 511.org website to get help with what public transportation was available. They entered the address of Jerry's apartment as the starting point, and the address of the zoo as the ending point. The Web site offered them a detailed itinerary, telling Julia exactly what to do. They printed this out; Julia felt confident that she'd be able to do it by herself.

In the morning, Jerry rushed to the office early for a meeting. Julia made herself an omelet and phoned her parents from Jerry's landline. She checked her email on Jerry's computer, packed a bag with a water bottle and some fruit, and set off to find the zoo. All the buses and trains were on schedule, and she arrived there without a problem. Julia was thrilled to be in the city. She thought about a friend in London, whom she would have loved to talk to right now, but she couldn't figure out how, other than to use a pay phone, and she knew that would be ridiculously expensive. Jerry had asked her to call and let him know that she had arrived safely. After a bit of searching she found a pay phone and called him. He didn't pick up, as he was having coffee with his boss, but she left him a message letting him know that everything had gone well. Julia bought the entrance ticket and got a map for the zoo. She wanted a cup of coffee, so she looked at the zoo map and found a restaurant. While having coffee, she looked over the map and planned out her trip inside the zoo.

First she saw the birds, then the monkeys. She went to the big cats exhibit and then stood for a long time watching a zebra eating. There was a theater at the zoo that offered daily "wildlife theater." The next show would begin in fifteen minutes. She was looking for the theater location on her map, trying to figure out how to reach it from where she was. She was proceeding in one direction, looking at the signposts along the side of the pathways. After a few minutes, she realized that the other route would probably be shorter. She got a bit lost and flustered, and arrived at the theater after the show had started.

Nonetheless, Julia had a fantastic trip to the zoo. She spent a long time watching the tigers lounging around in the shade and took a bunch of pictures with her digital camera that she would later upload to her Picasa account so that her friends and family could check them out. She enjoyed the California sunshine, and ate her lunch outside. She was still feeling very proud of herself for finding the zoo all by herself in a foreign country. Now it was 5 o'clock, and the zoo was about to close, but it was too early to head back to Jerry's apartment. He had said that he'd probably be home around 7:30.

Julia decided to try to find the Golden Gate Bridge. It seemed like a daunting task, however, and Julia really wasn't quite sure how to get there. She had no Internet access or else she would have logged on to the 511 Web site and printed out an itinerary similar to last night. When she finally found a pay phone, she dialed Jerry's number again. Luckily, he answered and she asked if he could look up directions from the 511 Web site from the zoo to the Golden Gate Bridge. Jerry was having a very busy day at work, and really didn't have the time to dictate the directions to Julia over the phone. Nor did he have much faith or interest in public transportation.

"Take a cab," he said, "It will be way easier." Jerry explained.

"Is there a phonebook there?" he asked.

"Yes," said Julia.

"Turn to the back pages and look up 'Taxi,'" said Jerry.

The pages of the phone book were old and tattered; many pages were missing. Julia found a company called Yellow Cab.

"You've got to call them, tell them where you are, and they'll come pick you up," said Jerry.

Julia hung up and called a cab. She waited about 10 minutes, and the cab showed up in front of the zoo. The cab driver was very friendly, and chatted with Julia while driving her towards the bridge. At her request, he dropped her off at the north end of the bridge, and then charged her what seemed like an arm and a leg for the ride.

Julia was very excited to see the bridge in-person, after seeing it in photographs ever since childhood. It was such a magnificent feat of engineering excellence! She walked from the north end of the bridge back toward the city, stopping for a while in the middle to admire the bay, the water underneath the bridge, the sailboats, and the view of San Francisco. It was sunny and windy, and there were a lot of other tourists and bikers enjoying the view. She took tons of pictures. Then she again found herself looking for a pay phone and discovered one near the restroom on the south side of the bridge.

"Here I am at the bridge!" she said to Jerry. "It's so beautiful!"

"Nice," said Jerry. "I wish I were there. I get off work in a few minutes. Have you figured out how to get home from there?"

"Well, I've got my itinerary," answered Julia.

"Yeah, but that was from the zoo. Now you're at the bridge. Do you know how to get back?"

"Oh, shoot," said Julia. "I didn't even think about that! Oh, man, this might get complicated. I guess I'll have to call the cab again, and they'll take me to the train station."

"How 'bout this," said Jerry. "I'll pick you up and we can drive into Sausalito for dinner. It's right on the other side of the bridge, and there are some fantastic places to eat there."

Julia was thrilled.

"So, it might be an hour before I get there. Wait close to the restrooms and the pay phones at 7:30. I should be there by then."

Julia was very happy to spend another hour in that beautiful spot, and quite relieved that she didn't have to worry about taking public transportation back to Jerry's house. Julia returned near the restrooms at 7:30 and waited. Fifteen minutes went by, but there was no sign of Jerry. Julia had no way of knowing that he was stuck in traffic. Finally he showed up around 8:00. They were both starving. After a bit of searching, they found a sushi restaurant in Sausalito and then took a stroll along the beach. Afterwards, they drove back to his home.

Virtual Companion

It is a few years in the future. In a very short time, Web on-the-go has become a universal phenomenon. Pretty much every global citizen has at least one Mobile Internet Device. Wireless access is available just about everywhere on planet Earth. Julia brought her Samsung MID to America with her, and also her Sony Pocket PC. She uses the MID to read emails and do some quick work online and the Pocket PC to write emails, read e-books, and more. The MID fits into her shirt pocket while the Pocket PC is in her handbag.

When Julia arrived in San Francisco, she pulled out her MID and sent a quick text message to Jerry, who immediately responded, letting her know he'd be waiting out front in his car, and then another to her parents back in the UK to let them know she'd arrived safely. Later on that night, Julia and Jerry were looking at the San Francisco Website and then the 511.org site to get clear on Julia's public transportation route to the zoo. On the 511.org site, a link was provided to download a transit trip planner application to a Mobile Internet Device. Julia downloaded and installed that application in her MID.

The next morning, Jerry rushed to his office for the staff meeting. Julia got up and began preparing her breakfast. At work, in the meeting, all members of the firm were offering updates on their current projects. Jerry gave his update early and then listened to the updates from members of teams in different departments. He had his laptop in front of him and launched the Google Latitude application, which showed that Julia was still in Palo Alto. He sent a short text message, "Hi, are you still at home?" Julia responded that she was eating her breakfast. Jerry texted back: "Have a great trip to the zoo!"

Around 9 a. m., Julia left the apartment and launched the 511 transit trip planner application on her Mobile Internet Device. The MID picked up her location from GPS and responded with a map showing her a map of the neighborhood she was standing in, including names of surrounding streets. The trip planner application prompted her to say or enter the place where she wanted to go. Julia said, "San Francisco Zoo." But the application couldn't recognize her British accent, so she had to enter the destination using the keyboard. Then the application asked her the next question, "When do you want to go? She entered "Now." Then the application came up with an itinerary, based on her current location, the current time, and current traffic conditions. This is known as "Information in Real Time."

Now, Julia's MID showed her a map with the directions that she needed to follow. It also had voice instructions. "Walk to that corner." The device picked up the GPS location and also did some internal calculations to arrive at Julia's new location. Then it said, "Turn right and walk to that bus stop." Julia reached the bus stop. The MID said, "It will take another five minutes for the bus to arrive. Have a dollar and 25 cents for the bus fare. It will be a ten-minute bus journey to reach the Caltrain station." After about five minutes, the bus arrived. Julia paid the fare and within a few minutes, reached the station.

Julia was very excited about her first bus and train trip in California. She was proud of herself for doing it all alone. In fact, she didn't really feel that she was alone. The voice instructions from the MID 511 transit planner application made her feel safe, and strangely enough, as if she had company. Meanwhile, Jerry was still in his weekly staff meeting, listening to people give their updates, while off and on checking the Google Latitude application on his laptop. He could see that Julia had reached the Caltrain station. In some cases, for obvious reasons, one would want to turn on the Google Latitude application's privacy feature, in order to block anyone (or specific people) from being able to see their location. In this case, however, Jerry and Julia were not at all concerned with privacy. Jerry really just wanted to know that Julia was doing okay. Soon the meeting was over and Jerry rushed to his cube. He called Julia to talk to her about her trip so far. Meanwhile, Julia's Mobile Internet Device told her that the train for San Francisco would be arriving at the platform shortly, and was giving her all the relevant instructions.

"I can't believe how easy this is!" Julia said to Jerry. That eased Jerry's mind, and he got back to work, not so worried that Julia might run into trouble finding the zoo. Julia boarded the train and began traveling north toward San Francisco. Meanwhile, her Mobile Internet Device continued to pick up her current location from GPS and informed Julia about each approaching station. When the train approached Millbrae, her MID let her know that she needed to get off at that stop. Thus Julia followed the step-by-step voice instructions and reached the San Francisco Zoo safely. Since Jerry and Julia were both online, every so often they would exchange small messages, making jokes and enjoying each other's virtual "company."

Julia reached the zoo and bought her entrance ticket. At the entrance, Jerry had let her know that there were instructions for installing the Zoo Guide application onto her Mobile Internet Device. Julia asked about that feature at the counter. The assistant told her that this application included the zoo maps, timings for shows, and specific directions to any location within the zoo, such as the theater and restaurants. Julia noticed that on every lamppost in the zoo, there was a wireless access point.

"What's up with the access points on the lampposts?" Julia asked the assistant.

The employee told her it was the Zoo Wireless Network and that the Zoo Guide application installed on her Mobile Internet Device accesses those points. Through the same application, Julia learned that she could also access the Internet. However, she had noticed already that she had coverage throughout the city, thanks to the new municipal Wi-Fi service in San Francisco. Julia went to the Zoo Guide Application Download booth; bringing her MID close to the counter, a Bluetooth connection was established between her MID and the counter; she then easily downloaded the Zoo Guide to her MID.

A message popped up: "Do you want to install the Zoo Guide application on your Mobile Internet Device? Yes or No." Julia selected "Yes," and after a moment another message popped up: "Zoo Guide application is successfully installed on your Mobile Internet Device." Julia now launched the Zoo Guide application, which immediately established a wireless connection with the nearest lamppost. Each lamppost access point had a unique identification name, and with that name, the application could access the current location of the user inside the zoo. Julia selected the option: "Locate restroom." Her MID immediately gave directions on how to reach the nearest restroom from the location where she was standing at that moment. Julia wanted a cup of coffee, and so asked the MID where the closest restaurant was. Immediately, it gave specific directions on how to reach the restaurant. While she was drinking her coffee, she explored the Zoo Guide application on her MID. She let Jerry know that she had downloaded the Zoo Guide application through the chat feature on her MID. Then Julia alerted her MID that she wanted to view the tigers; the Zoo Guide application gave her step-by-step instructions on how to reach the tiger cages. It also alerted her to the other animals she'd see on the way. Her MID offered Julia rich background information about tigers in their natural habitat, and about this particular tiger and its family. Julia really felt that she was walking with a real person, a guide, telling her all about the animals at the zoo. She took a few pictures of the tiger with her Mobile Internet Device camera. Those pictures were automatically uploaded to her Picasa account. Immediately, the tiger pictures were available in the public folder on her Picasa account. Right then, Julia noticed that her dad was also online. He was in London. She started chatting with him and sharing her pictures from the San Francisco Zoo.

Then Julia was looking at the wildlife show times in the Zoo Guide application. The next show would begin in 10 minutes, and she wanted to go. This time the Zoo Guide application on the MID gave her step-by-step instructions about how to reach the theater from her current location, and thus she arrived on time. During the show, she noticed that an elderly Chinese lady also had a Mobile Internet Device and was watching the show. But the elderly lady was listening to the show commentary from her MID.

Julia was curious to know what she was listening to. When she asked, the Chinese lady said that she was listening to the Chinese translation of commentary about the show. Once the show was over, Julia continued happily wandering about the zoo. She was confident that she would never get lost with that device.

Again, in the evening, she decided to go to the Golden Gate Bridge. This time she had no regrets that she hadn't planned ahead for this side trip before leaving home. When she left the zoo, Julia launched the 511 trip planner application. The application registered her current location from the GPS sensor and said, "Rightnow you are in San Francisco; where would you like to go?" Julia entered "Golden Gate Bridge." The 511 trip planner application came up with information about the public transportation that would take her to the bridge, and Julia arrived there safely, in a very short time. Almost the whole day, Jerry was watching her going from place to place on his laptop by running the Google Latitude application.

Jerry sent her the message: "It looks like you are going to the Golden Gate Bridge. Stay there and I will pick you up." Julia was happily walking on the bridge. This time, even without a cell phone, she easily stayed in touch with Jerry in the U.S. and her family in the UK. There was no need for her to use the pay phone or wait around near the restroom so that Jerry could find her. Checking the Latitude application on her Mobile Internet Device, Julia noticed that Jerry had left Mountain View and was driving north on Highway 101. After an hour, he parked his car and got out his Mobile Internet Device to select walking directions to reach Julia. The device was pointing toward the middle of the bridge. Jerry walked and noticed in his device that Julia was walking toward him. Both of them saw a handshake sign on their devices and by that time, they had bumped into each other. They spent some time together on the bridge, then searched for a good restaurant close by, using their Mobile Internet Devices. The result showed quite a few in Sausalito. They spent an incredibly enjoyable evening, and then headed back to Jerry's apartment.

These days, pretty much everyone has a cell phone. We pay a premium fee for the service, but much of the time we are traveling to places where our cell phone no longer serves us. Perhaps we are tourists in another country, or even just out driving or camping in a rural area. The technological advancements suggested in the second part of this story demonstrate a much more efficient method of wireless communication. The San Francisco Zoo Guide application could be applied to any zoo or theme park in the world. A similar downloadable program could be offered at Disneyland, the

Singapore Zoo, Great America, and more. 511.org is a wonderful concept for public transportation. They have already built the entire infrastructure necessary to evolve into the downloadable application referred to here. The next step would be to make this program work on a Mobile Internet Device. A further feature might include the ability to track an individual's current location from GPS. It is real-time information, real-time decision-making, with the ability to incorporate fluctuating variables such as traffic into consideration.

STRANDED IN THE WOODS

A True Story

Every so often, the unthinkable comes to pass. A small decision that we make in one moment meets with twists of fate seemingly beyond our control and suddenly all circumstances collide in a major, life-changing event. We hear of fabulous tales involving romance and riches that unfolded in this way, and then we hear of tragedies and disasters that similarly began as insignificant decisions on an afternoon much like any other. The tragic, true story of James Kim and his family began in this way, on a family vacation and road trip in the Pacific Northwest in 2006. James Kim, along with his wife Kati and their two daughters Penelope (four years old) and Sabine (seven months), had spent their Thanksgiving holiday in Seattle, Washington. After Thanksgiving, the family began the long drive back to their home in San Francisco. On Saturday, November 25, 2006, the family left Portland, Oregon, and drove towards the Tu Tu Tun Lodge near Gold Beach, Oregon. However, the family missed their turnoff from Interstate 5 to Route 42, a well-traveled road leading to the Oregon Coast. James and Kati checked their highway map and decided upon an alternate route, as opposed to turning around and finding the missed exit. The road they chose was called Bear Camp Road and bordered the Wild Rogue Wilderness, an extremely remote area in the southwestern part of the state.

"It's a narrow, winding mountain road with very few pull offs," said Patty Burel, a spokeswoman for the Rogue River-Siskiyou National Forest. "It's only one lane, so if two cars approach each other, one might have to back up to find a pull-off so the other can pass." Bear Camp Road is infamous to locals as a forty-mile stretch of twists, turns, and scary driving, even in the middle of August. Due to difficult terrain, lack of maintenance, steep drop-offs, and bad weather, the road is seldom traveled between the months of October and April. Though James' family passed three prominent signs as they continued up the road warning: "Bear Camp Road May Be Blocked by Snow Drifts," Mrs. Kim later told the police that they had noticed only one of these signs. Indeed, after running into very heavy snow higher up on the road, the family mistakenly turned onto a dirt logging road. A sign on the left-hand fork at this junction pointed drivers heading on Bear Camp Road to the left in order to continue on toward the coast. But the family turned right instead, onto BLM 34-4-38, driving for twenty-one miles before stopping because they were low on gas. At this point the family built a campfire out of old magazines and wood. Later, they even burned the tires from the car to help signal rescuers.

The gate on this particular road had been open, which allowed the Kims to mistake it for a continuation of Bear Camp Road. Mr. Kim's father claims that locals have said accidental detours at the particular junction are common. Though the Bureau of Land Management (who supervises the logging road) had rules saying that the gate should be kept closed, it is thought that it was left open by BLM employees in order to prevent locking in local hunters who could have traveled beyond it. On December 2, after six days in the same spot without any sign of rescue, subsisting on baby food, jelly, and bottled water, Kati and James pondered the road map again. They thought that Galice would be the nearest town, and that it should be located about four miles from where they were stuck. James Kim left his family to go in search of help. He was wearing tennis shoes, light clothing, and a jacket. He promised Kati that he would turn back the very same day if his effort to find help failed. However, he did not return. Search efforts for the family had begun on November 30, after co- workers of James Kim filed a missing persons report with the San Francisco Police Department. Soon, investigators discovered that the family had used their credit card to pay for a meal at a local restaurant. Shortly after this, a rescue operation was underway, including local and state police, over eighty civilian volunteers, the Oregon Army National Guard, and helicopters hired by James' father. The rescue teams spent several days searching for the lost family, but encountered no sign of them. The Kims did have a cell phone with them; however, they were out of range at the remote location where they were stuck. Eric Fuqua and Noah Pugsley, engineers from Edge Wireless, contacted the search and rescue party and offered to assist with rescue efforts. Thus on December 2, the same day James left the car in search of help, Eric and Noah began searching the data logs of cell sites in the hopes of finding records of repeaters that might have connected Kim's cell phone.

A mobile device like the cell phone carried by the Kim family constantly alerts cell towers and mobile switching centers as to its current location, but only when the phone is within range. The switching centers use this information to route incoming calls and messages to the tower that is closest to the user. In a situation where the cell phone is out of range, most often the location of the device is erased from the switching center. Sometimes switching centers do store recent communication between a device and the switching center for a 24-hour period. In the case of a missing person, or in this case family, Noah and Eric knew that it was possible that the switching center might still have the last recorded location of the Kims' phone before it went out of range or died. And in the event that the family had received a call or text message on their phone, that would make it much easier to find them, as these records are kept for billing purposes. Lo and behold, they found what they were looking for. On November 26, at 1:30 a.m., one of the family's phones had picked up a signal long enough to receive a text or voice message. The connection did not last long enough for the family to get the message or call for help. But it did allow Noah and Eric to retrieve this information and determine a 26-mile radius in which the family was probably located, narrowing the search to Bear Camp Road. Using this critical piece of information, on the afternoon of December 4, a local helicopter pilot found Mrs. Kim and her daughters walking along a remote road. He radioed the location of the family to authorities, and the three were soon airlifted to safety. Mr. Kim, however, had not yet been found. On Wednesday, December 6, at 12:03 p.m James Kim's body was found lying in one to two feet of icy water in Big Windy Creek. He had walked 16.2 miles from his car and was only one mile from Black Bar Lodge, closed for the winter, but fully stocked. An autopsy revealed hypothermia as the cause of death.

In the end, it was cellular technology that proved most valuable in locating the missing family. GPS is a fantastic resource for navigation and helping people find their way in certain circumstances, but it does very little to help in a situation like this, where the family was stranded without a signal on their cell phone. Without a signal, there was no way to track them with GPS. Tracking devices exist that send beacons to rescuers, but these are used mostly by backcountry hikers and skiers, and are rarely carried by people on a road trip. Similarly, satellite-based tracking technology does exist, but very few people want to have their location tracked constantly in the event that a rare disaster such as this were to take place.

Indeed, in this situation, cell phone technology served as a valuable lifeline for Kati and her daughters. However, this story also demonstrates the current limitations of cell phones and the network that supports them. For example, the battery life of these phones is short. Some batteries last a few hours, while others might last a few days. And though network coverage has expanded rapidly, there are still major gaps in coverage, especially in remote areas. And these areas, for obvious reasons, tend to be the places where coverage for emergency purposes is most important.

James' father has made clear his frustration that cell phone records storing the last transmission to James' cell phone on the night they were stranded were not investigated by the police. Indeed, the records were not found at all until two engineers from Edge Wireless volunteered their time. He has brought up other factors that prevented a speedier rescue that might have spared the life of his son. He stated that neither the hotel where they had stayed, nor their credit card company, would release information about the family to him due to privacy laws. He claimed that media planes disrupted rescue efforts and that general search-and-rescue efforts were confusing and lacked order.

In the summer of 2008, the Bureau of Land Management and the Forest Service installed six large signs along Bear Camp Road, an information kiosk, and mile markers from the Galice access to Gold Beach. The new signs are meant to clearly mark the route from Galice to Gold Beach and also to let drivers know that the road is not maintained from November to May, when snow drifts along the route are a real possibility.

James Kim remains well-known for his work as a television personality on the international cable network TechTV. He had worked more recently before his death, however, as a senior editor at CNET.

Out of respect for the parties involved, I will not replay the Kim family tragedy in the following hypothetical scenario. I will instead use fictional characters and events in order to portray how Web on-the-go technology has the potential to save lives and avoid disastrous events such as the ones recounted above. My hope is that this information will address some of the key limitations of the cellular technology that indeed helped to locate the Kims, and offer viable, more advanced technological solutions to ensure that this type of catastrophe is not repeated in the future.

Jerry and Julia are now married, with two children, Jonathan, who is 13, and Nancy, who is 8. Recently the family bought a Samoyed puppy that is just over a year old and named him Snowflake. The family lives on a quiet street in Palo Alto, next to an older couple named Wilson and Martha. Wilson and

Martha enjoy the children, and so Jerry and Julia end up spending quite a bit of time at their house. Wilson tells incredible stories from his time serving in the Korean War. Martha is an accomplished cook and used to sing in a popular choir in San Francisco. She teaches Julia some of her best tricks in the kitchen and when asked, sometimes sings songs for the family. Usually, however, she refuses, muttering that her voice just isn't what it used to be.

This year, Jerry and Julia decide to take a road trip to Seattle for Thanksgiving. Jerry's sister lives there, and they haven't been up to visit in a year and a half. Knowing the dog can't come on the trip, Jerry asks Wilson and Martha if they could watch him for a week. Wilson is actually allergic to dogs, but wanting to be a good neighbor, he agrees to let the dog stay with them while the family takes a vacation.

"Now you be careful," advises Wilson. "Remember what happened to that family in 2006."

"Oh, don't worry about us," said Jerry. "Technology has actually improved a lot since then. Each of us will have our Mobile Internet Device with us, and our minivan even has one installed. This alone would prevent that tragedy from occurring again."

In all honesty, Wilson had no idea what they were talking about. In his 80s, he had just gotten his first computer, and needed constant lessons and supervision just to check his email. But he smiled and wished them well. Jerry made sure to leave him the phone number of their sister, just in case anything went wrong.

Jerry's family had a fantastic time in the Pacific Northwest with his sister. They even traveled briefly into Canada to check out the city of Vancouver. It rained every day, but they had been expecting this and didn't let it affect their plans.

Back at home, Snowflake was having a grand old time with Wilson and Martha, even though Wilson greatly disliked living with a dog. Martha loved Snowflake, and walked him twice a day around the neighborhood. Wilson did not complain, but secretly kept his eye on the calendar, counting down the days until Jerry and his family would come home to retrieve the dog.

After Thanksgiving, the family headed back south on I5. They stopped in Roseburg, Oregon, for dinner at a diner, and then continued on.

Unbeknownst to them, they were actually searching for the same road that the Kims had been looking for, Highway 42, which would lead out to the coast. They had reservations at a hotel in Gold Beach.

In the backseat, Nancy and Jonathan had their MIDs out and were playing games. As children are on long road trips, they begin to bicker for no reason. Within a few minutes, their bickering transformed into a full-fledged argument and Jonathon was pinching Nancy's arm, hard.

"Children, stop arguing," demanded Julia firmly. "It makes it difficult for your father to concentrate on the road."

"But he's PINCHING me!" screamed Nancy. By now she was crying.

"Well then, stop being so annoying!" yelled Jonathan.

"KIDS!" shouted Jerry. "STOP FIGHTING RIGHT NOW!"

But Nancy was bawling and Jonathan pinched her again in the arm.

"All right," said Jerry. "I'm pulling over so we can sort this out."

Jerry got off I5 at the next exit. He pulled over and reprimanded the kids. Jerry thought he'd turn around away and get right back on the highway. However, the road they were on was extremely narrow, and he could not find any place suitable to make a U-turn. It was evening, and still raining, and driving in reverse was out of the question. Jerry drove for ten minutes without finding a place to turn around. He checked GPS, but the road popped up as "unnamed." Julia checked her Mobile Internet Device. It showed the road and the GPS location, but the map simply showed forest surrounding them, and offered no other details about the route.

The kids, excited by the twisting road, stopped fighting and began looking out the window, though they couldn't see much in the dark. Jerry decided to keep driving, assuming that at some point soon he would encounter a crossroads or at least a house with a driveway. However, the further they drove, the more narrow and winding the road became. And to make things hairier, suddenly the rain turned to snow.

"Wow," said Julia. "We must have gained some elevation. There's gotta be something up ahead. Let's go a little bit further."

Thirty minutes later, they had still found no place suitable to turn around. The snow drifts were piled high, sometimes making it difficult to drive.

"I think we have to stop here," said Jerry. "Hopefully, the weather will clear by the morning, and we can find a way to turn around."

They kept the van's heater on throughout the night, and surprisingly, they all got a great night's sleep. When they woke up in the morning, they could not believe their eyes. It was a winter wonderland. Snow on the branches of the evergreens made them droop down low. The sun was shining, and glinting off the white blanket surrounding them. They had parked in the middle of the road. On one side of them was a steep drop-off and on the other side, an open meadow. Their van, they realized, was helplessly stuck. Snow had piled behind them and in front of them on the road, and driving through it was impossible. It was hard to see where the road stopped and the meadow began.

"I have to pee," said Nancy. "Me too," said Jonathan.

Trying to keep things light, and stay on the bright side, Jerry said, "Well, lucky us! We've got the luxury of going anywhere we want. There's no one to see us out here." He pointed toward some bushes close by. The kids got their rain boots and jackets on and trudged out of the van. They were so excited to be in the snow.

While the kids were out of the minivan, Jerry and Julia exchanged glances.

"This isn't good," Jerry said. "But I think our best bet is to sit tight here. If we try to go find help, who knows what might happen."

"I agree," Julia replied. "If we conserve, we've got enough food for about four days, and we've got warm clothes. I think we'll be fine."

Jerry pulled out his MID and checked for a signal: nothing. He set the device in the sun to keep it powered up, and Julia did the same. Jerry was grateful that these devices were now solar-powered. With his technological background, he knew very well that these devices might prove to be the family's lifeline.

That day the family gathered wood for a fire, and tamped down an area in the snow to sit around it. Jerry and Julia were surprised at how quickly their worries disappeared. Jerry stopped fretting about work when he realized that there was absolutely no way he could do anything about the high-priority items that needed attending to at the firm. They all realized, of course, the direness of the situation. However, here they were in a stunningly beautiful place. As long as they imagined that they were on a grand camping adventure, the kids were having the best time of their lives. They built snowmen and sprinkled bottled juice over snow to form ice cream. There was no homework or school. They wished there were a stream to fish in, or some wild berries.

Meanwhile, back home, poor Wilson was getting frustrated, and a little worried that the family had not called to tell him they would be late. They were now two days behind schedule. He was also agitated that Snowflake and Martha were becoming such good friends. He called Jerry's sister, but no one picked up. Not knowing what else to do, he called 911, and let them know that the family was missing. The police came to Wilson's house and asked him dozens of questions about Jerry and his family, none of which Wilson could answer. He didn't know where Jerry worked, where the kids went to school, the names of friends or relatives. But he did remember the conversation with Jerry the day they left.

"He did say something about a mobile device he was taking that he was sure would help in case of an emergency," Wilson told the police. "Jerry works with computers and stuff. He's really on top of all the new technology."

This was all the police needed to hear. They had heard that some police departments were using the Google Latitude application with their work, and wondered if Jerry might have this installed on his MID. They put a call into Google and asked for their help in locating Jerry. A couple of officers went down to Mountain View to meet with Google engineers. They searched their database by entering in the GPS coordinates of his house and his first name: Jerry. Almost immediately, they came up with the last connected location of Jerry's MID, off the highway on a small road near Roseburg, Oregon, a few days ago. Indeed, Jerry had Google Latitude on his MID.

"Wow! That's incredible!" the policeman said. "We're very grateful to you. But if you don't mind me asking, how is it that this is open information? I mean, you could just track anyone who has this application."

"Well, there is a privacy option," one of the Google engineers explained, "but it's different from cell phones. With a cell phone, it's up to the cellular network whether to make this information public or keep it private. In the case of Google Latitude, each user decides whether to keep their location public or private. In this case, I'm wondering if Jerry didn't keep his location public just in case of an emergency."

"Well," said an officer. "This is really an important step forward in technology. I'm going to talk to my boss and let him know how valuable this program would be for our police force. I mean, we could have access to that same information, right?"

"Absolutely," said the engineer. "As long as someone is keeping their information public, and they have the Google Latitude application, you'll be able to find them."

The officers thanked the engineer profusely, shook hands with them, and then began the hard work of rescuing Jerry and his family. A search-and-rescue team was organized. It was decided that helicopters would fly toward the last registered location of Jerry's MID.

Knowing that the family had at least three MIDs with them, and that all of those devices were solar-powered, they decided upon an innovative approach to the rescue efforts: Instead of using their eyes or powerful cameras to find the family, they attached a high powered wireless access point with an extremely strong Wi-Fi signal to each helicopter, and then fixed an antenna on each chopper, pointed toward the ground.

Meanwhile, back at the minivan, it was daytime. Jerry was napping, Julia was cleaning up, and the kids were playing games on their MIDs. All of a sudden, a message popped up: "Open Wi-Fi connection available." Nancy yelled, "Daddy, there's Wi-Fi!"

Meanwhile, Jonathan selected the Wi-Fi and accepted the connection. In one of the helicopters, an officer said over his radio: "It looks like they're here!" From the helicopter, a wireless connection was established to the device. Immediately, a chat application was launched.

"Hi, are you OK?" from the helicopter. "Yes, we are," Jonathan typed.

Nancy yelled:"If you say we're ok, then they won't come and get us! Tell them we're doing really bad!"

Jonathan said, "Don't worry. They will come." "Are you Jerry?" came from the chopper.

"I'm Jonathan, Jerry's son. Dad's taking a nap." "Oh, I see. Who else is there?"

"My mother, she's folding our clothes and cleaning up the van and my sister is playing with her games on the MID."

"Looks like the whole family is having fun."

"Yes, we are. We like the fresh air, fresh water, singing birds, and today the weather is also good; nice and sunny."

"OK, then, enjoy your vacation. We'll let you be."

"NO! Please come here and take us away. Our van is stuck."

"I was kidding. We're coming closer. Do you see the handshake sign?" Jerry and his family quickly packed their things and prepared to leave this beautiful spot. The helicopter lifted them to safety and took them straight to the hospital, though there wasn't any reason for them to be there, as they were all in good health. The entire rescue, from the 911 call to the airlift, had taken less than twenty four hours.

After these incidents, local authorities, including the U.S Forest Service, U.S Bureau of Land Management, police and fire services all had a meeting. They decided to install Wi-Fi infrastructure in the woods along the road where Jerry's family had gotten stuck. There were no cell phone companies operating in that location, and there was no competition regarding the licensed spectrum versus the unlicensed spectrum. So the authorities were free to provide powerful Wi-Fi signals.

They were ready to push forward this agenda, but when local environmentalists caught wind of what was being planned, there was an uproar. They claimed that installing Wi-Fi in the woods would take away from their experience of spending time in nature.

Some avid campers stated in local newspapers that wild places were meant to stay wild, meant to remain as places to go where one really could leave it all behind, which of course included the Internet. Environmentalists planned a rally, and even a few city council members showed up to oppose the Wi-Fi in the woods plan.

Responding to the uproar, the authorities once again came together to discuss the plan. After debating the issue hotly for several hours, they come to a consensus. They decided to make the wireless system in the woods a Local Area Network. This meant that everyone who was traveling in the wild area could be in touch with each other, but they could not be in touch with anything outside the network. Hikers could seek assistance from a ranger in case of an emergency, for instance, and rangers would easily be able to locate anyone within that area who was in need of help. But because the system was localized, no one could access the Internet. You would not be able to watch YouTube videos while backpacking in the forest, for example. This seemed like an appropriate and helpful use of the technology and quelled most of the criticism revolving around this very contentious issue. Once these wireless stations were built, if someone entered into this remote area, his MID immediately picked up the wireless signal and a pop-up appeared on the screen signifying that a Localized Area Network connection was available. A warning message also came up: "You are entering the woods. This is an area of heavy snowfall and road closures. If you need more information, please establish this connection."

Once the connection was established, the browser was launched. The current road conditions were visible. If you got stuck in the snow, it was possible to click on the SOS link for help. If someone clicked the SOS link, the local authorities received an alert and responded immediately via chat or voice. Now that they knew the location of the people in trouble, they could find them easily, with a helicopter, if needed.

All of the wireless stations in the woods were solar-powered, and non-obtrusive. In fact, no one ever noticed them. During the summer months, people felt comfortable planning camping trips in the area, and the National Forest was being used, respectfully, more than ever before.

Hikers found another way to use the wireless network. They formed groups; each group took a different route to a single location. Using the localized version of the Friend Finder application, they were able to keep track of each other. In addition, the voice communication over the Local Area Network helped them stay connected. If one group saw a bear in the forest, they immediately took pictures and sent them to the other group, which had walking directions to them and got there in 15 minutes.

All thanks to wireless in the woods.

SHIPPING CONTAINERS

Once the kids had grown up and were enrolled in school, Julia had plenty of free time on her hands. She decided to open up a fashion boutique selling dresses, shoes, and accessories; this had been her lifelong dream. Years ago, Julia received her undergraduate degree in fashion design but never really used her training after moving to America. She had many good friends in the fashion design industry, both in America and in the UK, who excelled at designing dresses, shoes, handbags, and more.

Julia imagined that not only would she really enjoy introducing new designs to the market, but that she would be good at it. She opened her store on a bustling street corner in San Francisco. Within the course of a few years, her boutique's reputation grew; many local celebrities had become faithful customers.

The fashion designs for her products were done in the U.S. She documented the process with a step-by-step guide describing the finer details of manufacturing, material selection, and color, then sent the design documents to manufacturers in China. In two weeks, her Chinese partners created a sample and returned it via FedEx. Once Julia approved the sample, the clothing article was manufactured in bulk, packed into large boxes, and loaded into shipping containers that would then make the journey to America.

The containers were loaded onto an oceangoing container ship that was headed for the Oakland port. When the shipment arrived in Oakland, Julia's agent Mike cleared the containers with customs and Homeland Security, loaded the containers in a truck, and delivered them to Julia's warehouse in San Francisco. Within a week, Julia put the new items on display in her shop (and on her Website) and began selling the products to her customers.

As she became acquainted with the business world, Julia found that everything went quite smoothly, except for the shipment of the containers.

Michael Fernando, or Mike, was Julia's shipping agent. He had been in the shipping business for more than twenty-five years. His business involved making sure that the containers belonging to his client were picked up safely from the Chinese manufacturer in a truck, then loaded on a ship in the Chinese harbor with all of the proper documentation. When the ship reached the Oakland port, Mike made sure that the containers cleared customs, then loaded them onto a truck and delivered them to the proper client and warehouse. Finally, he took the empty containers and dropped them off at a lot in Alameda.

Julia always had problems with Mike; however, she never fully realized that the challenges she faced in receiving shipments from China did not begin with Mike. In fact, it was the global system of shipping containers across the world that was mired in problems. Mike had an uphill battle when trying to deliver these containers to his clients efficiently. Let's delve into some of these challenges.

When Julia was just getting started with her business, she decided to have a grand opening for her shop. On opening day, she wanted to display different styles of dresses, handbags, and accessories. Never having owned her own business, there had been a steep learning curve for Julia. In the months building up to the grand opening, Julia burned the midnight oil with her designers and collected a myriad of new styles for men, women, and children. She sent the designs to China after identifying manufacturers and placing her orders with them. The manufacturers that she chose, located in the northeastern region of China, committed to the project and set a firm delivery date for the goods.

Julia confirmed all this with Mike, who assured her that he would take care of picking up the materials from the manufacturers in China and shipping them to the U.S. Mike said it should be very simple. He couldn't foresee any problems, though he did tell her that in this business, things rarely go exactly according to plan.

Since Julia had ordered such a large quantity of materials, Mike said, they would probably fill up an entire container. He walked her through the process. His first order of business would be to arrange a container in China. Once the goods were ready for pick-up in China, he would lease a container from XYZ Container International (one of the world's largest leasers of transportation equipment). Then Mike would ask his business associates in China for a truck, and hire a driver to pick up the cargo from the suppliers and load it all into the container.

"Once the cargo is collected from all your suppliers," Mike assured Julia, "the truck driver will drive to the Shanghai Harbor port. The container will be loaded onto an ocean carrier. On average, it takes eleven days for the ship to arrive at the Oakland port. Then the container will sit in customs from a minimum of four days to a max of ten days, during the busy seasons, such as Christmas."

Things were shaping up nicely. Julia had a commitment date from her manufacturers and a pretty good idea about the amount of time it would take for the goods to arrive at her warehouse. After speaking with Mike, she decided it was safe to fix a date for the grand opening of her fashion shop. Julia talked to a celebrity friend and requested that the celebrity help her in opening her dream store. The celebrity agreed to be present on that day, modeling some of the fashions and signing autographs. Giddy with excitement, Julia went ahead and printed the invitations, and started distributing them to all her friends and professional contacts. She also placed a giant advertisement in the local newspaper.

As the big day drew closer, Julia grew increasingly tense and restless. Each morning, she sent an email to her suppliers, and each afternoon they promptly replied that everything was on schedule. Before she had placed the order with those particular suppliers, Julia had consulted myriad references, as well as the supplier's list of satisfied customers. Everyone had offered positive feedback,including comments regarding the quality of the product and the timely delivery schedule. On the delivery date, the supplier sent her an invoice and let her know that the goods were ready to pick up.

Julia breathed a huge sigh of relief. She was satisfied that things seemed to be on schedule. Julia thanked her suppliers and assured them that she would return to do business with them in the future. She added that if the product was as lovely as the sample they had sent, she would not hesitate to recommend them to her friends and other professionals in the fashion design industry.

Julia phoned Mike and informed him that the goods were ready to be picked up in China. He promised that he would talk to his business associates there as soon as possible and initiate the shipping process. For the first time in months, Julia relaxed. She spent the weekend with her family and looked forward to the big opening day.

On Monday morning, Julia went into her office. She had somewhat urgent emails from the suppliers waiting in her inbox, asking politely but firmly when she was planning to pick up the goods. Julia called Mike to figure out what the hold-up was. Mike told her that he had run into a problem leasing a container. Perturbed, Julia hung up the phone and briskly drove to Mike's office in Oakland. When she got there, she found Mike looking frantic, speaking on the phone. When he hung up, Julia did not try to hide her annoyance.

"Mike, what's the problem?" asked Julia. "I've got to keep things on schedule. I've planned a huge event that is entirely dependent on this shipment showing up on time."

"Calm down," replied Mike. "I have talked to several companies around the world and I've figured it out. By today or tomorrow, I'm almost positive that I'll find a container for shipment."

"Why, what's the problem?" asked Julia. "I see tons of these containers on trucks on Highway 880. Even on my way here, just around that corner, I saw a huge depot stacked full of containers!"

"In the U.S., we have plenty of containers," Mike explained, trying to remain even, "but in China, it's actually surprisingly difficult to locate an empty container."

"Why?"

"Well, as you know, these days the whole world is importing from China," Mike answered. "All the containers arrive in the U.S. filled with goods, and then, once emptied, they get stuck here and in other importing countries. So in China, there is always more demand for empty containers."

"Why don't we just send all these empty containers back to China?" Julia asked.

Mike replied, "It would just cost as much to send empty containers back as it does to ship them to the U.S. full of cargo."

"Well, Mike," replied Julia, "I'm not here to try to solve big global issues. I've got my shop opening in a few weeks. Tell me how you're going to get me my goods in time."

Mike sighed, exasperated. "I talked to a friend of mine in Singapore. He said he'll help me find an empty container in China through his contacts. I'll let you know how this pans out by tonight. I'm not worried, though. I think we'll be fine."

That evening, Mike called Julia to confirm that his friend in Singapore was able to help him, and that the goods should arrive in time for the opening of the shop. Mike also found a truck driver to collect the empty container and pick up the goods from Julia's suppliers. That night Julia slept well. In the morning, she received emails from her suppliers telling her that the goods had been picked up by Mike's agency. Julia was happy and thanked him for making a special effort to get this done. Then she asked him again whether he thought the goods would arrive on time. For obvious reasons, she was concerned. Mike assured her that they would definitely arrive on time.

He said that it would take another ten to fifteen days for the container to get from Shanghai to Oakland. Julia had allowed a week of cushion in case of any unexpected delay. Because of the container issue, she had already lost two days of her seven-day cushion. The schedule was getting tight now, and this only added to Julia's stress level.

Julia felt embarrassed, calling Mike every morning to check the status of her cargo. So after two days, she called him again to get a report. He told her that because the Chinese were currently celebrating a holiday with large festivals everywhere, the container had not been loaded onto the ship till that day, and that the Shanghai port was busy because of the backlog. He hoped that the container would easily clear Chinese customs, and be loaded today.

Julia lost her cool and yelled at Mike over the phone: "I hired you to do this job and do it right!" she shouted.

"Don't expect my business in the future!"

The next day, Mike called her and told her that the container was now on the ship, which was set to begin its journey at any moment. After a week, Julia called Mike to see about the progress

of the container. Mike told her it was in the middle of the Pacific Ocean and would arrive soon.

Julia was amazed at how archaic the entire system seemed. "Isn't there a way to track the container movement, like FedEx does?"

"Unfortunately for us, no," said Mike. "In the case of FedEx, from the source to destination, everything is handled by one company. This makes it much easier for them to collect the current location information and the status of the goods. But in this case, our container is leased from one company, the truck drivers are small business owners, the Port Authorities are a government agency, ocean carriers are their own company. Everyone has information on your container, but unfortunately, the information is not integrated and made available at one location. This is one of the biggest challenges I face in this business."

"Well" replied Julia, "just this once, can you please give me an honest answer as to when my goods will arrive?"

"To be frank with you," replied Mike, "I don't know. And there isn't anyone involved, or anyone from any other company who could commit to one specific date of arrival. In this business, we use the term ETA, or estimated time of arrival."

Mike went on to explain other issues that made his job difficult: "If you're going to be in this business, there are definitely other things you need to know about the shipping industry. Now,

don't you worry about this, because the weather has been good, but sometimes containers even fall from the ships that carry them, usually during storms. It is estimated that over 10,000 containers are lost at sea each year. For instance, on November 30, 2006, a container washed ashore on the Outer Banks of North Carolina, along with thousands of bags of its cargo of Dorrito's Chips."

"Yikes," said Julia.

"Containers lost at sea do not necessarily sink," continued Mike, "but rarely float very high out of the water, making them a difficult-to- spot shipping hazard."

It was actually pretty interesting. Julia would have enjoyed listening to Mike's talk, had it been someone else's goods that were late. But because it was her goods that were stranded out in the Pacific, with only an ETA and her shop opening up very soon, she couldn't enjoy the conversation. With her containers still on the ocean, Julia came to the conclusion that she should postpone the opening of her shop. She immediately called the local newspaper and asked them to cancel the advertisement. She phoned her celebrity friend and explained the problem And called all her friends to let them know that she was canceling the grand opening. Once the goods arrived, she would have a simple opening ceremony.

In fact, Julia's goods reached the Oakland Port a week after the original grand opening date. It took five days for customs and Homeland Security authorities to clear the goods in Oakland. Then Mike hired the truck that picked up the container from the port and delivered it to Julia's warehouse. It took a week for Julia's team to unload the goods from the container. Then, Mike sent the truck again to pick up the empty container. Julia asked the truck driver where this empty container would go. The truck driver told her that he was going to dump it at the container depot in Alameda. Julia now felt guilty that she herself had added one more container to the pile, lying idle in the U.S. like a heap of trash.

With a simple ceremony, Julia indeed opened up her shop. The dresses she had ordered pleased her greatly. They were very wellmade, some of them even hand-embroidered, in rich colors and at a very affordable price. Julia was extremely happy with her business.

Though she had a bit of trouble getting things started, since the shipping incident, everything had been going quite smoothly. After a few months, Julia decided to expand her business. She wanted to have a section for shoes and leather materials. She identified a reliable, good-quality, fashion shoe manufacturer in China; however, they demanded that she order a huge quantity of shoes—too many for her small store. Julia had made many friends in the fashion business. One of her friends, Sandra, had a fashion shop in San Diego. Julia and Sandra joined forces and each placed orders with the Chinese shoe manufacturer. They decided they would split the goods fifty-fifty.

Though she thought long and hard about it, Julia decided to use Mike's business again. This time, she knew that the shoes might be delayed, and yet she wasn't so worried, as there was no deadline to meet. She told Mike just to give her a call once he delivered the goods to her warehouse.

Julia also told him that 50 percent of the goods would have to be transported to San Diego. Mike said he would take care of that, too. He wasn't sure what the cheapest way to ship the shoes would be, so he came up with a few potential scenarios, but in the end decided that it would be cheapest to split the load in two while still in China, have one load sent to L.A., and one load is sent to Oakland. Mike would hire a truck to bring the cargo from L.A. to San Diego.

The truck drivers were instructed to deliver the goods to the warehouse. Julia's team unloaded the packages and called Mike when they were done to pick up the empty container. Another empty container had been added to the depot in Alameda. Similarly, Sandra's team also unloaded the goods, and then informed Mike. Mike sent the truck drivers to pick up the empty container and dump it in the yard in L.A.

The following day, in San Francisco, Julia's team moved the packages to the shop and started putting shoes in the showroom. Similarly, in San Diego, Sandra's team was busy putting the shoes out on the floor. All of a sudden, Julia realized that every single one of the shoes was for the right foot. There were no pairs! She called Sandra, and, sure enough, she had only shoes for left foot.

Angry, Julia again phoned Mike.

"The supplier might have done it to prevent theft," said Mike.

"Another problem with this industry is that things are often stolen along the way. However, if there are no pairs of shoes, then they won't get stolen."

Taking the blame, Mike offered to pick up the goods from San Diego at no charge and move them to San Francisco. Then, Julia's team would pair the shoes and send the remainder back to San Diego. Again, the process was far from smooth. But in the end, both Julia and Sandra sold a great many of these shoes, and made a fabulous profit.

Let's imagine a few years have passed. These days almost everyone has their own Mobile Internet Device. With applications such as Google Latitude, people can remain connected to their friends all the time. Of course, anyone who wants privacy can choose the privacy option on their MID. Along with this new technology, an idea among interested parties has begun to form. What if we extend these same features of MIDs to the shipping container business? What if we treat each container like a human being carrying an MID? Similar to a Friend Finder application, we could have a Container Finder application.

Mobile Internet Devices are emerging as extremely powerful and affordable devices. It was thought that if MIDs could be installed on all shipping containers, and fitted with a customized applications like Google Latitude that would capture the movement of these containers, the container shipping industry would be drastically improved in a very short amount of time. In addition, wireless infrastructure was available almost everywhere—in the city, at the port, and on freight ships.

Let's assume that everyone in the shipping industry agreed to this proposal. Soon, the International Organization for Standardization (ISO) came up with new specifications for containers in order to accommodate Mobile Internet Devices. The ISO consulted the military in order to get this done quickly and efficiently. The military had already done the grunt work in creating electronic devices that work in extremely tough environments and conditions.

Since the Mobile Internet Device is almost like a computer, other modifications were devised as well. Since security of the cargo is a major concern, MID would control the door and lock of each container. Thus in addition to the normal physical lock, an electronic lock was created for each container. For a power source, both battery and solar power panels were provided to ensure that a container's wireless device would never fail. Once the specifications were finalized, the container manufacturing industry started releasing these new "intelligent" containers for trial.

Now let us return to Julia and Mike. Julia's shop has become a well-known landmark in the city; fashion-savvy residents and tourists flock to her store to see what new fashions Julia has. Both shop and online sales are on the rise. Julia has a list of extremely reliable suppliers that have never yet let her down. The only challenging aspect of her business has been in shipping goods from China to the U.S. Almost every day, Julia receives complaints from her warehouse workers about Mike. There was always some new disaster.

But for the last few months, Julia noticed less chaos and confusion in her shop and in the warehouse because of shipping concerns. Surprisingly, the shipping process (the backbone of her business and once the bane of her existence) seems to have become flawless. Complaints from her employees at the warehouse have all but stopped. Just to make sure that things really are going smoothly, Julia called the warehouse team members one day and asked about Mike. They told her that in the last few months, Mike had consistently delivered on time, and on the date specified. Not only that, they reported, but he'd drastically reduced his fees.

Julia couldn't believe it. She asked her warehouse workers jokingly: "Hey guys, why didn't you tell me? Do you think you are only allowed to bring me bad news?"

Julia phoned Mike to make sure that he was in the office and asked if he had a half hour to meet with her. He made space in his schedule. Julia drove to Jack London Square. As she entered Mike's office, she noticed a lot of changes. Last time she had visited, and every visit up to this point, the office had been a confusing whirlwind of shouting employees, complaints coming in from all sides about lost containers, and lines of customers questioning employees angrily about the status of their cargo. Julia noticed that new computers had been installed, and that almost all of Mike's truck drivers had an MID in hand as they entered and exited the office.

Mike invited Julia into his personal office. He poured her a cup of coffee and then Julia said: "Mike, we used to yell at you when things were not going smoothly. But my guys have informed me that for the last few months, things have been really different. They told me that you are delivering cargo on time, and that your charges are also considerably less now. First of all, I want to thank you for your wonderful service. It's not fair on our part, if we just ignore your quality service and take things for granted. I don't want you only to hear from me when there's a catastrophe. So, Mike, thanks again."

"You're welcome," said Mike.

"But I am so curious to know what has changed," continued Julia. "After hearing about all those issues in the shipping container industry, I thought things were always going to be difficult. My curiosity is piqued."

Mike beamed, obviously proud. "Do you have time now?

It might take a while to go into the details and I won't be able to explain everything. I would need to show you some things on the computer."

"I've got all the time in the world," Julia replied.

Mike gave Julia a short introduction to the Intelligent Containers with embedded MIDs, and the Container Finder that closely resembled the Friends Finder application.

Then Mike demonstrated with real information on his computer. "Most of your suppliers are in the northeastern part of China," he told Julia. He showed her the map and the location. "Now, let's assume that we need to pick up goods from your suppliers. I'm going to look for containers in that region within a fifty-mile radius."

The map showed ten to twelve small container icons, some green, and some yellow. Mike clicked on a green container and it had complete details. This container would be available in a day or two.

"So if I needed to, right now, I could reserve this container for shipment. If I reserve it, then this container is mine. Once the container becomes empty, the MID embedded in the container will send me, as well as the truck driver on my service, an email notification that it's ready for pick-up."

Julia was truly excited and asked about the yellow icons.

"That means the full container is not available," replied Mike, "but that a part of that container is available. If the desired destination of goods matches, then we can reserve that, too. Let's see…"

Mike clicked on the yellow icon, and saw these details: "Shipment of dresses to Chicago, via Oakland port. If we didn't have a full load, then we could use that container," explained Mike. "When it reaches Oakland, we could take our stuff out and the container would then be put on a railroad headed for Chicago."

Julia expressed her joy: "Wow, so we don't necessarily need to bring one full container all the way to the U.S. and then leave it in the yard lying idle."

Mike said, "Since I've begun using these partial load containers, I have been able to reduce my charges."

Julia said, "That's really fantastic! But I have a question. When you unload our goods at the Oakland port, it seems like the Chicago-bound goods are at risk of theft, like you mentioned before, aren't they?"

Mike said, "That's a good question," and went on to explain. "Now all the shipping documents are electronic. There is no paper. This yellow container is with one supplier right now. That supplier is loading his goods bound for Chicago. Once the supplier loads the goods, he needs to load the shipping documents to the Mobile Internet Device on that container. It will include supplier details, importer details, details about the package, invoice, etc. Then the supplier locks the door, using his electronic authentication details. Most of the time, the container is connected to the Internet using the city wireless or the satellite wireless connection. Periodically, it sends information such as its present location, rate of movement, and shipping documents to the servers at the back end."

"Then the truck driver picks up the container. The truck driver cannot open the container, because he doesn't know the authentication information. He takes the container to your supplier. I will send the authentication information to your supplier. Using that, your supplier loads your goods and the shipping documents. Then he locks it."

"Then the truck driver takes the container to the port. The port authorities take their device close to the container. Their device makes a connection to the MID in the container and they immediately get all the details about the movement of that container to various suppliers' locations and also the shipping documents of all the goods. Once cleared, the container is now loaded onto the ship."

"So there is a log of events maintained inside the MID in the container, keeping track of the opening and closing of the lock, and the identification details of each person who opens the container. If anything is missing, it will be easily traceable. So theft is very much reduced. Because of that, nowadays the suppliers are confident that goods will not be lost in transit. They don't do shipments of shoes in two containers, with all the left shoes separated from the right shoes. It's just not necessary."

Julia asked, "Can anyone watch the movement of my goods?"

"We can set the visibility," replied Mike. "Most of the time, once you lease the containers, they are visible only to us and to the shipping agents. Besides us, the containers are always visible to the owners and to government authorities. Since the containers are visible now, it's easy for us to track the movement. Do you want to see your shipment that left China last week?"

Mike showed her the map with satellite mode. Julia could see the Pacific Ocean. Then he located the ocean carrier ship and zoomed in further, to a huge stack of containers sitting on the ship. Julia couldn't believe it. She could actually see her container in the middle of the Pacific!

"You can set the alerts, too," continued Mike proudly. "My guys and I like to receive an alert when the ship enters the San Francisco Bay. Once we receive the alert, my drivers will head to the Oakland port, even taking into account the traffic on I-880, so they will be there right on time."

Julia asked, "What about the delay by the Customs authorities and Homeland Security authorities?"

Mike said, "Their jobs are so much easier now. They carry a scanner that looks like a gun, point that scanner at the ship, and get a complete list of all the containers on that ship. Then they randomly choose a few containers and look for the details. The authorities have access to complete shipping details. They even can track the complete itinerary of any container."

"Nowadays," continued Mike, "the Customs authorities can begin their part of the process when the ship arrives at the port, and can continue checking the containers even when they are being unloaded from the ship. They finish up when the container is placed on the truck. So it really is very fast."

"In fact, now government authorities get alerts from the Chinese authorities when the ship leaves a Chinese port. So the authorities in the U.S. can monitor a ship's movement and its container stack throughout its journey, right from its source."

Julia was overwhelmed by all this new information. She thanked him and went home, feeling totally confident about her business. Since every single one of her business issues was now resolved, Julia could begin spending more time at home. However, on her drive home today, she began dreaming of expanding her business by opening more shops in other parts of the country. She decided she would delegate the responsibility of inland movement of the containers and goods to all her shops in various parts of the country to Mike. Together, Julia and Mike would help each other to grow their individual businesses.

In conclusion, I offer some detailed technical notes regarding the new Intelligent Containers. For the movement of some highly sensitive goods, the government authorities might prefer to monitor total movement, directly from the source all the way to its destination. In order to accomplish this, government authorities might form a global organization called Global Security for Containers (GSforC). When a company is involved in shipping sensitive goods, it needs to procure the key from the GSforC organization. With that key, the supplier opens the container and loads the sensitive goods, and then the supplier locks the container and returns the key to the GSforC organization. It's a completely electronic key, so it's easy to exchange throughout the network. While in transport, no one can open the containers. When the goods reach their destination, the importers seek the key from the GSforC organization.

Port Authorities, including Chinese Customs, U.S. Customs, and Homeland Security seek the key from the GSforC organization and open the container for any investigation they would like to make. Once they are done with their search, they need to return the key to the GSforC. With this simple procedure, nearly all security threats from global terrorism at ports are eliminated.

Since all containers are now embedded with MIDs, these devices are now capable of sending email alerts. If a container falls into the sea, the drifting container can now send an email alert to the concerned authorities (such as the supplier, importer, ocean carriers, container owners, or insurance companies) for help. Because of this feature, the ocean is no longer polluted with falling and floating containers.

On the ship, these containers are stacked. The containers at the bottom of the stack cannot make an Internet connection via the satellite connection. To resolve this, all the containers in the ships form an ad-hoc network, a network connection with their neighboring containers on the sides, top, and the bottom. Then the containers in the top row make a connection to the Internet using satellite communications. In this way, all the containers can comfortably communicate their current location and current status to their servers at the backend during their journey on the ocean. Once the goods are delivered, these containers will be stacked in the depot, like the one in Alameda. Usually, the depot authorities check the conditions of these containers after every trip. If they find any damage, they submit their quote to the owner. The owner or their insurance agency has to approve those quotes. The depot will do the repair work. Once it is done, the owner will make the payment.

In the container industry, this M&R maintenance and repair service is a very complicated and cumbersome business process. But, with the embedded MID, this is resolved, too. At the depot, these containers are stacked and again form an ad-hoc network. Now they can communicate to their business owners over the Internet, through the city wireless or through satellite communication. After inspection, the authorities submit their quote to the MID in the container itself; then the MID sends the quote to its owner to approve. Now the containers can send alerts to the depot authorities to go ahead with the repairs. Once the repair is completed, the MID in the container informs its owner about its health condition. The MID tracks all repairs made to it in the course of its lifetime also. Using a combination of the MID, the Container Finder, and the Wireless Infrastructure, a simple, effective solution to the global issues of the shipping containers is within reach. We have all the technology needed to create a worldwide transformation of this challenging and critical global business.

INTELLIGENT VEHICLES

Jerry had been working hard. He hadn't taken a real vacation since they had become stranded in their van in the Pacific Northwest. When summer rolled around, Jerry told his family that it was time for everyone to take a break. Julia's business was now well-established and running smoothly. She agreed that the timing seemed right. The kids were enjoying their summer holidays, but after a month away from school, they were bouncing off the walls and needed a change of scenery. After a long discussion, everyone agreed that a trip to Disneyland and southern California sounded relaxing, fun, and easy to plan.

They decided to drive to Southern California in the minivan. Both Jerry and Julia would take turns behind the wheel. They packed their luggage into the van and left home in the early morning. The kids were still sleeping and Jerry and Julia carried them to the car, where they continued to doze. Julia also soon fell asleep. Jerry was driving and listening to country music on the radio. Every time he passed through a city, he would lose his country station and have to start scanning for another one. Luckily, most of the time, finding another station that he liked was easy.

After two hours on the highway, the kids woke up and asked for breakfast. They found a McDonald's right away and pulled over. After breakfast, both Julia and Jerry picked up cups of coffee from Starbucks. When they started driving again, Julia took her turn behind the wheel and Jerry tried to take a nap. The kids were asking for their favorite radio stations. Julia tried to explain to them that the radio station signals from the Bay Area were not available any more because they were now two hours away from home.

Jonathan muttered, "There must be a better way to do it."

Julia tried playing some CDs that the kids liked but they complained, saying that they had already listened to those CDs so many times and were bored with them. With this kind of racket going on, there was no way Jerry could sleep. He knew that if the kids were not calm, then Julia would not be able to drive peacefully. So Jerry started telling stories to keep the kids engaged.

"You know, guys," Jerry began, "when I was a kid, my dad had some old car. It worked great, but it was old, even for that time. His car didn't even have a radio! So my father always carried a transistor radio with him when he went on a trip. He took that small radio pretty much everywhere: in the car, to the beach, to his construction job. In those days, there were definitely no iPods and not even car radios."

Luckily for Jerry and Julia, the kids were listening to him attentively, obviously curious. To keep the story (and the relative peace) going, Jerry began to speak about his grandfather. He didn't tell the kids, but he was making up this part of the story.

"In my grandfather's time," continued Jerry, "they had one radio, a very big one, and only at home. It was so huge, you could not take it out of the house. So when they were away from home and on the road, there was no entertainment, no music, and no talk shows."

By now, even Julia was interested in this part of the story, and also relieved that the kids were engaged. "Really!?" exclaimed Julia. "Then what did they do when they were on the road?"

Jerry was rolling with it, and couldn't believe that he actually had an answer to this. "Just like those Sound of Music kids," he continued, smiling. "Everyone sang while they were traveling. Do-re-mi-fa-so-la-ti-do! Doesn't that sound kind of cool? I'm guessing it would have been a lot of fun. Everyone sang along. There were no radio commercials, no bad news to listen to. Unlike nowadays, when a family travels together, like right now, everyone is either listening to the radio, or their iPod, or something, but rarely do they talk to each other, or sing songs on the way."

Intrigued, the kids agreed to sing some songs. Julia joined in. These songs were like a lullaby for Jerry and he immediately went to sleep. After singing for a while, the kids became quiet. They looked out the windows at the passing farms, trees, road signs, other cars, and faraway villages. After a while, the kids went to sleep again. Julia was enjoying this quiet time. She liked driving. Most of the time, in the city, she couldn't even reach the speed limit. Now, on the highway, she was cruising along at the California maximum, 75 miles an hour. Even so, many other cars kept passing her.

She realized that they must be going 100 miles an hour. Slowly, she started hitting the gas pedal and the speedometer crept up: 80, 85, 90, 95, and finally the car reached 100 miles an hour. Julia enjoyed the thrill of going so fast. She assumed that it was OK to go 100 miles an hour on the highway. Since everybody around her was doing it, she decided that it must not be a crime.

Suddenly, in the rearview mirror, she noticed a highway patrol car with its lights on, getting closer to her and finally right on her tail. Julia pulled over to the side of the road. When the car stopped, everyone woke up. Quickly she made sure that everyone was wearing their seat belts and also that the kids were buckled in.

The cop approached the driver side window and asked to see her driver's license, registration, and insurance papers. She opened her purse and found her driver's license while Jerry searched through the glove box in the car to find the registration and insurance papers. The minivan was registered in Julia's name, so Jerry didn't know anything about the documents pertaining to the car. As he got it out of the glove box, he noticed that the registration renewal was due. Julia had completed the renewal online, but the smog test was still pending, so the registration was not complete. The family insured both cars on one plan, so Jerry knew that the insurance was up-to-date, but he saw that the papers in the car were old and expired. Fortunately, Jerry kept a small card with proof of current insurance in his wallet. In the end, the officer was satisfied with the documents and just gave Julia a ticket for speeding. However, he warned Julia that with kids on board, it was extremely dangerous to drive above the speed limit.

Julia took the ticket and apologized for speeding. She asked the officer, "Can I ask you a question? It wasn't only me. Almost everybody was driving 100 miles an hour. How come you came after my car?"

The officer said, "Maybe you're right, Ma'am. But usually, once someone sees a police car, they reduce their speed. There's no way I could catch them. You didn't slow down, and that's why I got you."

Julia went on and said, "But I saw a sign back there that said the speed limit is controlled by radar surveillance."

The officer replied, "We're working on this, but in all honesty, the surveillance system is not that effective."

Julia was pretty upset with the ticket. Jerry comforted her with soft words and asked if she wouldn't rather spend some time in the backseat with the kids. They traded places and Jerry began, once again, driving toward L.A.

In the evening they reached Los Angeles. Jerry and Julia had full faith in their van's built-in GPS system. They were totally confident traveling to new places without buying printed maps and they never bothered to find driving directions online. Usually, the GPS system served them well, but this time, things were not going according to plan.

The Holiday Inn in Anaheim had recently moved to a new location whose street details were not available in the van's GPS navigation system. The minivan was nine years old, and so was the GPS navigation map. Jerry had been thinking about upgrading the system with new maps, but he hadn't yet done so. Similarly, the GPS navigation system in the car had plenty of details about restaurants and attractions, but many were out-of-date.

Jerry reached Anaheim and called the hotel for directions. On the phone, they told him which exits and streets to take. However, it was late in the evening at this point, and Jerry had been driving for many hours. He was exhausted and couldn't follow the directions very well. After a while, he got the sense that he had been driving around in circles. They spent nearly an hour searching, when the children finally saw the neon Holiday Inn sign on top of a building.

Finally they reached the Holiday Inn, had their dinner, and got a good night's sleep. In the morning, they had breakfast in a nearby restaurant and then drove to Disneyland. As they parked the car in the parking lot, Jerry asked the kids to remember the sign closest to where they had parked. They were in the Minnie section. In Disneyland, all the signs in the parking lot were names of cartoon characters: Mickey, Minnie, Simba, and so forth.

They spent the entire day inside Disneyland, going on several rides and watching the parade. In the late evening, they watched the fireworks. It was nearing eleven o'clock when the family finally found their way back to the parking lot. The kids were so tired that they began to fall asleep. Julia sat them down on a bench and they immediately leaned against her shoulder and closed their eyes.

Since Jerry had asked the kids to remember the location of the car, he hadn't paid any attention to it. They racked their brains, but couldn't remember. Finally, they decided to wake up the kids and ask them. Groggy, the kids opened their eyes. But after seeing so many Disney characters, they couldn't remember, either.

Jerry asked Julia to stay on the bench with the kids while he searched for the car. He checked every level of the lot, looking for something familiar. After spending a good amount of time, Jerry located the minivan and picked up his family. Exhausted, they made their way back to the hotel.

They had learned their lesson, and for the next few days, they took the hotel shuttle to Disneyland. They had to wait a while for the shuttle each day, but at least they didn't have to worry about parking. Since Jerry didn't have to drive, the whole family was having fun together, even on the shuttle bus.

After having a fairly relaxing time in L.A. they headed south to San Diego. Jerry was driving. At one point, they had to pass through a toll booth. Julia suggested taking the Fastrak lane, since they had a Fastrak device in their minivan. The kids asked Julia what Fastrak was.

"Well, this little device in the minivan communicates with the toll gate machine. We have a Fastrak account with money in it. Because we have already paid, they simply deduct the toll fee from our account."

They reached San Diego and found their hotel, then went to the beach and had a fantastic time splashing in the waves. Soon after that, the kids were hungry and it was time for dinner. They decided to drive downtown. As they entered San Diego proper, Jerry and Julia began looking for a parking space. It was Saturday afternoon, and people were out and about. There were no parking spots to be found. After twenty minutes of searching, they finally found a spot, but unfortunately it was for a compact car, a much smaller car than they were driving. With their minivan, at least a foot of the van would be in the red zone. Jerry was tired of circling around looking for parking, however, and decided to take the risk. He squeezed the minivan into the tiny parking space. Then they went to a Mexican restaurant and had a very good dinner. When they came back, Jerry saw a ticket on the windshield of the van, for parking in the red zone. The ticket, however, mentioned that he could send in an explanation if he thought that the charge was not valid. He could also contest the ticket in court.

They spent two days in San Diego visiting the Zoo, Sea World and of course the beach, whenever they had some spare time. Today they were getting ready to check out of their hotel and begin their drive back to the Bay Area. They decided they were up for one final adventure, to visit the San Diego

Wild Animal Park. At this park, the animals were roaming free, not locked in cages, and people had to travel around in vehicles to see the animals. The kids were so excited about this, they couldn't stop fantasizing.

Jerry was checking out at the hotel counter and asked the employee if she knew anything about the Wild Animal Park. The woman talked enthusiastically about it, adding that she had just visited there with her grandkids. Jerry asked her for directions. At that point, a kind-looking man, also checking out of the hotel, smiled at Jerry and told him that his family was planning a trip to the park that day. He introduced himself as Bob. Bob told Jerry that he knew how to get to the park and that Jerry could just follow his car. Jerry thanked Bob and decided this was probably his best bet.

The two cars entered the highway. Bob was a bit of a reckless driver, and was accustomed to driving fast. He just assumed that Jerry would be able to follow him. Bob was switching from lane to lane and Jerry found it so difficult to follow Bob, they eventually lost sight of him. With a smile, Julia reminded Jerry about the episode of "Seinfeld" when George and his girlfriend Susan drove the car so fast, Jerry and Elaine couldn't keep an eye on George's car and lost sight of it.

Meanwhile, the kids were looking outside, reading the signs. "Dad!" yelled Nancy. "The sign said that the next exit is for the San Diego Wild Animal Park!" So they took the exit and followed the road signs until they found the park. They spent their afternoon checking out all the wild animals, and later in the evening, started their return journey back to the Bay Area.

After a completely satisfying and adventurous vacation, they were back home. Later on that week, Jerry sent a letter to the San Diego city authorities explaining to them that his car was well within the parking space and not in the red zone, as he had been charged with. They didn't agree, however, and in the end he had to pay the fine.

These days, almost everyone has a Mobile Internet Device. Not only that, but most of the cars being manufactured are equipped with these devices. In Julia's new 2010 minivan, every seat has its own MID. All the devices inside the car form a network. In addition, through the antenna on the roof, the devices in the car are able to connect to any wireless access point or any satellite Internet communications system. Almost every city now has the infrastructure for wireless support. And along the highway, wireless access for the Internet is nearly always available.

With these new circumstances, let us reimagine Jerry's family vacation to L.A. and San Diego. In the early morning, the family packed up the minivan and left for L.A. Jerry was driving. Julia and the kids were sleeping in the back seats. Jerry was listening to music, but he was no longer dependent on local radio stations, and so was no longer scanning around for a new station every hour. Nowadays Jerry listens to the radio on the Web. With the Mobile Internet Devices installed in his minivan, paired with his ability to access the wireless Web from anywhere on the road, Jerry was able to listen to his favorite songs continuously while his family slept.

As Jerry drove south, his wireless connection was passed along a network of different wireless infrastructures. Some cities had built wireless access points in the lampposts lining the highway. In other places, he connected via satellite communication.

Jerry could listen to music from all over the world in his car. Though he loved country music, he began to branch out. This morning he found a station playing only music from South Africa; he was absolutely entranced. After about an hour of driving, Jerry wanted to catch up with the morning's news. With the ability to tune into stations reporting news from around the globe, offering unique perspectives on issues like the current economic crisis, Jerry began to gain a real appreciation for the technological advancements that allowed him such freedom in accessing information. He felt connected to people on different continents, and was stunned by the stories he heard being reported from other countries. He knew that there was plenty to worry about, especially in the current economic downturn, but somehow, in tuning in to this global perspective, he felt a glimmer of hope that people would be able to get through these hard times.

Jerry knew that the kids would wake up soon and be asking for breakfast. Jerry didn't want to eat at McDonald's. He disliked fast food and didn't want his kids to get into the habit of eating it. Julia was just waking up, and softly he asked if she could find them a good place to eat breakfast. Using the Mobile Internet Device in the car, Julia selected their current location. The GPS system immediately identified where they were and showed a local map. Julia searched for local restaurants, found a few, and was even able to read reviews. She picked one praised for its healthy, local fare, got the driving directions, and helped Jerry navigate to the location. Jerry parked the car and the kids woke up. The restaurant was cozy and the food was delicious.

Nancy, the youngest, asked: "Dad, how did you know about this place?"

"My parents brought me here when I was a kid," he replied.

Nancy looked at him sideways and then sang, grinning: "Liar, liar, pants on fire."

Now Julia offered to drive. Jerry was in the back with the kids, who were using their Mobile Internet Devices in front of their seats. For a while, they were listening to their favorite music with head-phones. Then they were watching videos on YouTube, checking email, and chatting with friends. The family also had recently purchased Sling-Guide from Dish Network, a system that allowed them to find, watch, and record television programming from thousands of channels on their MIDs, among other things.

So the kids spent some time watching TV, too. Though Jerry and Julia had not done so, it was possible to set up limitations on these vehicle MIDs, similar to a home computer, to ensure that the kids weren't getting into any trouble online.

Jonathan saw his friends online and began chatting. Then he shouted, "Mummy, you know what? Seth is going to be at Disneyland tomorrow too! His family is driving to L.A. right now!"

The kids were totally occupied and keeping themselves busy. Julia was listening to romantic music, and Jerry easily drifted into a very deep sleep. As they weren't near any cities, the highway was empty. Julia was reaching the speed limit. She looked around the highway; there were no police in sight. She decided to hit the gas pedal. Slowly the speedometer crept up and up until it just passed the 100-mile-an hour mark. At that point she got an alert with a beeping sound in the Mobile Internet Device of her minivan. This was a new sound that Julia had never heard before. Jerry woke up.

Jerry said, "Julia, I think that means that you just got a ticket."

Julia reduced the speed to 75 mph and opened up that alert message on the screen of the MID. It read: To the driver of 4MAG700 vehicle: You have just exceeded the California speed limit of 75 miles per hour on highway I5 on March 25, 2010, at 12:34. Since this is the first ticket on your name for this calendar year, this is only a warning. However, if you exceed the speed limit again, you will be required to pay a fine of at least $150.

Julia was amazed at this new technology. She knew that her minivan was equipped with this system but, as a city driver in a new car, she had not yet

---

needed to be warned in this way. Julia was relieved and relaxed, and made sure to keep the minivan within the speed limit.

Julia asked Jerry: "How did they do that? How did they know that I was speeding? There was no highway patrol, no police, no siren... ."

"Well, It's sort of like a stopwatch," Jerry replied. "At one location, they capture the vehicle identification and the start time. Then at another location, they capture the end time. So they know the distance and the time taken to travel that distance. With that calculation, the system can deduce whether your vehicle was within the speed limit or if it exceeded the speed limit."

"The part that I most appreciate," Jerry continued, "is that the system refers to master data and looks for the number of tickets on your name. Since this is the first one for you, it happens to be a warning."

Julia said: "In the old days, radar surveillance wasn't effective. Remember the police said that during our previous trip? It seems as though now, with this wireless Web on-the-go, highway surveillance is extremely effective."

Julia was content with the way things had been going so far. She was enjoying driving on the highway, as this was her first road trip to L.A. At one point they hit some traffic and the freeway became clogged and slow. Suddenly Julia heard a loud noise. The car behind them had bumped into their rear bumper. Julia could see that it was a very old couple behind them. They both pulled over to the side of the highway. They didn't argue, as the circumstances were very clear, and agreed to exchange license information.

Julia took her Mobile Internet Device close to the car of the older couple, who also had MIDs installed. Both devices made a Bluetooth connection. Using their MIDs, they exchanged license and insurance information. In addition, the MID in the car had all the information about the car's insurance, so when prompted, it automatically sent an alert to the insurance company that there had been an accident, including details such as date, time, and location.

Julia said she wanted a break from driving, so Jerry took his turn at the wheel. Of course, this time they were no longer dependent upon ten-year-old maps in an antiquated GPS system. As they arrived in L.A. they had no issues locating their hotel. Embedded Mobile Internet Devices in the car made the navigation enjoyable. These days, the car received the information in real time, so Jerry and Julia could make intelligent decisions, taking into account current weather conditions and traffic reports.

Locating the car in the parking lot at Disneyland was no longer an issue, either. In fact, Jerry and his family didn't even bother to remember the parking location by using the Disney characters on the reference signs. Now both Jerry's car at home and the minivan were a part of Jerry's family network. The family members could easily locate each other using applications such as Google Latitude on any of the family's MIDs. Using his handheld Mobile Internet Device, Jerry could pinpoint the exact location of his minivan. As he walked toward the van, a hand-shaking sign appeared. Without any difficulty, the family found their van.

After having fun in L.A. for two days, the family drove to San Diego. Nowadays, they don't have that big Fastrak device sticking to the windshield. As the minivan got closer to the toll plaza, the Mobile Internet Device received the following message from the toll servers over the wireless connection: "Do you want to pay this $4.00 toll from your Fastrak account?" Jerry said "Yes," and that was it. Since the system is wireless and fast, people are now able to pay the toll before they reach the tollbooth, so there is no congestion at the toll gates.

Just like last time, in downtown San Diego, Jerry took the risk of parking the van in a small space meant for a compact car, with the rear of his van edging into the red zone. This time, however, the San Diego traffic officer on duty had a Mobile Internet Device, too. The traffic officer prepared the message for the parking ticket, and then took two pictures using the camera in her MID. She then established a Bluetooth connection with the MID in the car and exchanged the photos and parking violation information, giving the ticket to Jerry.

With this system in place, after giving the ticket to Jerry, the police immediately send all the parking violation information to the city server through wireless Internet. This means less confusing paperwork. The police don't need to update anything at their office. Since they simply attach the picture, there is seldom any reason for anyone to dispute the ticket. Most of the time, people just pay the fee.

After dinner, Jerry's family came back to the car. There was no paper ticket on the front windshield. But when he got into the car, the ticket popped up on the MID in the car. He saw the pictures, too, and knew he could not argue this case. He immediately paid the parking ticket over the wireless

Web. At the bottom of the ticket, the city police had a printed message: "Please access the city Website for details about available parking spaces at any time. We keep updating the parking details on a real-time basis." Jerry accessed the site and was amazed to see all the available parking spaces around downtown San Diego.

When Jerry and his family were leaving San Diego, and heading to the Wild Animal Park, Jerry again met Bob, who offered to show him the way to the park. Though Jerry knew he could easily follow directions to the park on his MID, he liked the idea of following Bob's car. This time, however, both Jerry's family and Bob's family met in the parking lot of the hotel. Jerry added Bob's car to their friend's network on the MID. This allowed Jerry's minivan to easily follow Bob's car. He could always see the car's current location on the minivan MID screen. The MID also registered the map in real-time and gave exact directions to reach Bob's car's current location.

It is absolutely invaluable to have sophisticated Mobile Internet Devices embedded in the car. And it is equally, if not more, important to provide ubiquitous wireless access in order for these devices to be widely used and enjoyed.

Up until this point it has been necessary to provide the background (and current) context for the situations that I have expounded upon. In the following chapters I will switch gears slightly and focus solely on the Web on-the-go concept and its various applications in our daily lives.

EDUCATIONAL INSTITUTIONS

It was a weekday evening. Everyone in the family had returned home and was engaged in the normal evening routine. Julia was preparing a healthy salmon dinner for the family and Jerry was helping her in the kitchen. Jonathan, a high school student, was doing his English homework upstairs in his room. Nancy, now a middle school student, was finished with her homework and was watching TV for a while. She kept flipping through channels, trying to find something interesting, but there really wasn't that much on.

She wanted to get online where she could watch the episodes she really wanted to watch, or chat with her friend Monica, or upload some photos to her Facebook page. However, Jonathan was using the computer to do his homework, and of course everyone knew that schoolwork was the priority. Suddenly, Nancy remembered her Mobile Internet Device. It was brand-new; her Dad had given it to her for her birthday. She was just beginning to explore all the possibilities of the little device. Excitedly, she ran upstairs to get it. Using the MID, she logged onto YouTube and began watching some "Arthur" clips. She still loved the Arthur kids show, and had been watching it ever since she was little. She didn't dare mention it at school, but at home she still watched "Arthur" whenever she could. Jerry had recently bought the entire family MIDs and they all stayed connected via Friend Finder. Using this application, they were able to see where other members of the family were throughout the day.

"Nancy," Julia called from the kitchen. "What are you doing in there?"

"I'm watching 'Arthur' on my MID," replied Nancy. "It's totally cool. Now I don't have to wait for Jonathan to be done with his homework! I can do whatever I want online right now!"

"Watching 'Arthur' on the MID?" Julia asked, incredulous.

"But that screen is small! It can't be good for your eyes to watch videos on that tiny screen."

"Well, I don't mind it," replied Nancy.

Jerry was listening to this conversation from the kitchen, where he was chopping up carrots for a salad, and suddenly he realized that what Julia said was probably very true. He didn't want his kids to grow up squinting and with bad eyesight. He decided that it was OK for the kids to use the MID in that way when they were away from home, but when they were using their MID at home, they shouldn't have to be staring at that tiny screen. He reminded himself that the MID was just as powerful as a regular computer, just as quick and effective, with the exact same computing and networking power. In other words, there was no reason why Nancy shouldn't use it just like a home computer, except for the fact that the screen was so small. An idea began to take shape. Because he was a tech guy, the new idea got him excited; he began to whistle while he chopped up the veggies.

Later on that evening, Jerry did some research. He used to have a docking station, a simple plug-in that allowed him to transform his laptop into a substitute for his desktop computer, yet without losing the mobility of the smaller machine. He searched for a similar docking station for the MID. Surprisingly, when he Googled this, so many products popped up from different vendors that he was amazed he hadn't heard about them before.

After a busy day at the office, Jerry stopped by the electronics shop on his way home. The employees at the store were extremely helpful and proudly showed off the new technology. Jerry had no problem finding exactly what he was looking for. He brought it home and after dinner, everyone gathered around to see how it worked. Nancy was especially eager to test out the new system. With Jonathan enrolled in Advanced Placement courses, he had been spending hours every evening in front of the computer doing his homework, leaving very little computer time for Nancy.

Jerry opened the box and plugged the MID into the docking station. Included in the package were a wireless keyboard and a wireless mouse. There were outlets available to connect to a computer monitor or television. With a cable, he connected Nancy's MID to the television and let Nancy take over. She was amazed at the new and improved system.

"Hmmmm," said Jonathan. "I think I might start doing my homework on the television. It certainly seems like I'd get more done with this new system."

Nancy shot him a dirty look. Only five minutes had gone by, and already Jonathan was plotting to take away her newfound technological perfection.

"Just kidding, sis," Jonathan said, laughing, but Nancy wasn't so sure.

As Jerry and Julia put dinner on the table, Nancy pulled up her favorite "Arthur" shows and happily settled in to watch. Plugging her MID into the TV was so simple, and now she didn't have to squint.

In the kitchen, Julia said to Jerry: "Thank you, sweetie, for doing that. I had no idea that was an option. It sure is amazing what's out there these days."

Nancy was enthralled with the new set-up. Watching "Arthur" on their large HDTV was so cool. The sound effects were amazing, and the image was clear. After a while, she decided to see what else she could do. She opened up Google Earth and typed in her home address, clicked on the satellite viewing mode. She typed-in the road where they had gotten stuck up in Oregon, and after a little difficulty, found the exact spot where they had been stranded. She called the family in to look. Next, she found their street and zoomed in so close she could see the terrace of their house, the minivan parked in the parking lot, and the small square of tilled earth in the backyard where they had a vegetable garden. Then she selected Street View and started "wandering" around the neighborhood. She was having so much fun.

Nancy looked at her mom and asked, "Do you want to see Grandpa's home in the UK?"

"Sure," answered Julia.

Nancy typed the UK address and got the satellite view of her grandparents' home. She clicked on the Street View option, but a message popped up that said the Street View for these locations was not currently available. Julia wondered why.

"Google employees fixed cameras on the roof of a small car," Jerry explained, "and went around the streets making videos. Then they processed these videos and created a Street View option. I saw those cars driving around the Bay Area. Perhaps they haven't done that yet in Europe."

Julia really wanted to see the Street View of her childhood home. "It seems like anyone could do that," Julia said, "I mean, take a look at YouTube. Anyone can upload a video. Why not allow people to do this for their neighborhood? I could ask my friends in the UK to go around the street with their video recorder and capture the street view, and then upload it to Google Earth. It seems like Google could save themselves time and effort by allowing people to do this."

"Hey," Jerry said, "that's not a bad idea. Living with a techie is rubbing off on you!"

Nancy remembered that one of her friends had told her about live video cams on the National Geographic Website. She logged on to a live streaming camera of an animal preserve in Africa. Elephants were crossing the TV screen in real time! This really blew Nancy's mind; maybe she could become a scientist who studies animals in Africa.

The next day, Nancy returned to school. During science class, she raised her hand before the teacher started the lesson and told everyone about the live streaming video from Africa, and about checking out different parts of the world on Google Earth, on their huge HDTV. Nancy explained how easy it was to hook her MID up to the TV, and how that made it possible for her family to use the Internet together.

Nancy was really good at school and was already beginning to imagine how this new technology could be used in the classroom. The science teacher, Mrs. Lampkin was really excited, as were the students. She could see no reason why she couldn't implement such a system in her classroom. She asked Nancy for permission to contact her father via email. After school that day, Mrs. Lampkin sent an email to Jerry, asking about the details of the docking stations for the MID. The teacher was interested in which system might be cost-effective and feasible for the school. She was already envisioning a system like this in each classroom.

Jerry made a few phone calls and got the opinion of some of his colleagues at work. After a few days, he responded to Mrs.Lampkin, suggesting a docking station that was both inexpensive and had received high ratings among professionals and ordinary citizens alike. This particular company offered an impressive discount to schools, since they were selling their product for educational purposes. Jerry was thrilled to be able to help out in this way; he also got in touch with one of Jonathan's favorite teachers at the high school, including him in this offer as well.

Web On-The-Go

On a trial basis, both Nancy's middle school and Jonathan's high school bought and installed a few docking stations for MIDs. All that was necessary to make the system work were a big flatscreen monitor, a keyboard, a mouse, an MID, and a docking station. There was no need for a functioning desktop or laptop computer system, or an operating system such as Windows or Mac OS. No extra software needed to be installed on the MID. For the students who chose to do so, the teachers began offering an MID homework option. The teacher would compile the homework for the week as an electronic file, and students could download this file onto their MID by plugging it into the docking station. As some kids did not have MIDs, there was always a hardcopy of the homework available as well. Those who got into the habit of putting their homework on the MID realized that their backpacks were much lighter now, and there was less confusion about exactly what they needed to bring home each night. When the students were done with their homework, they could transfer it to their teacher electronically. They were saving paper as well!

Nancy's science classroom was beginning a unit about the rainforest. Last Friday, Mrs. Lampkin had alerted the students about the unit, and had challenged them, as a pre-investigation, to come to class on Monday with five new pieces of knowledge about the rainforest. When she introduced the topic officially, she found that her students were extremely dialed-in to this subject matter already. One student plugged his MID into the docking station and showed the class a nonprofit group he had discovered, called the Pachamama Alliance, based in San Francisco, that was working with native tribes in Ecuador to protect their rainforest home. The teacher was very impressed. Another student showed a live video cam streaming in the Brazilian rainforest.

Mrs. Lampkin had never had a more successful introduction to a new topic. She realized at the end of the day that she herself had learned a wealth of new information and passed this success story on to all the teachers. Within a short time, the school had outfitted each classroom with a MID system. All the teachers came up with more and more inventive ways of implementing the technology. Mrs. Lampkin couldn't help but notice a new enthusiasm in staff meetings. Teachers were excited about teaching again. With all the stress about state standards and testing in recent years, she had noticed a frustration building among educators, stemming from a feeling of being limited by state-mandated expectations. With the new MID system, she realized that these standards could be met in a new way that actually excited both teachers and students. She was curious to see whether test scores would improve because of it, and she had a strong hunch that they would.

Because this system was implemented using MIDs, the schools didn't need to worry about constantly upgrading their computer infrastructure, and computer software. A new software system, known in the technical world as Software As A Service, was available. Schools needed only to provide the docking stations, a monitor, keyboard, and mouse, and these tools didn't require constant upgrading. In addition, the school needed to offer wireless infrastructure, which was not very expensive.

Jonathan, a busy high school student, also found that the MID was extremely useful for taking notes in class, submitting homework, and staying organized. In addition, many of the books in the high school curriculum were now available as e-books, so his backpack was much lighter these days. Using the docking station at home, as well as at school, Jonathan basically had his desktop computer with him all the time, and it was the size of his hand.

When a substitute teacher came to fill in at Jonathan's school, her job was much simpler. Each teacher has an online "sub folder," filled with interesting and relevant activities. Jonathan used to dread the days when his teachers were absent, because the subs never had a clue what was going on. Now, they simply plugged-in the classroom MID, and a full curriculum appeared on the screen, including bell schedules, school maps, and seating arrangements.

The students could see exactly what they had to do, and in what order. There was also a roster for each class, and an area to write down behavioral issues if needed. When the official classroom teacher returned, they found that the students had accomplished real work, and could easily see who had misbehaved.

A week ago, both Jerry and Julia attended the back-to-school night at Jonathan's high school. This was typically a challenging, though enjoyable, evening for them. They had to, in essence, follow the route that Jonathan took every day as he moved from class to class, visiting each of Jonathan's teachers in turn, and getting a sense for what the year had in store. However, the high school was huge, and Jerry and Julia didn't know their way around. Luckily, the school has an MID application for students and parents, including maps, schedules, teacher names, and contact info. This application can be used from home as well, so parents can remain actively engaged in their student's education. This time, when Jerry and Julia arrived at the high school with their MID, they saw a brief introductory note from the principal and a classroom map and schedule. It was actually fun!

Nowadays, computers are used extensively in the library. Almost all libraries provide their catalog on the Web, and usually people use this online catalog to locate books in the library. Card catalogs are a thing of the past, and most kids have no recollection of the days when library books were documented on small pieces of paper. You can search for a book based on the title, author, or other key words, or even book awards. You can manage your account online from home, reserve a book, request a hold, and renew already checked-out titles. Many libraries offer email reminder services, letting people know when books are due, when holds are ready for pick-up, and when fines are being accumulated. Libraries also provide Internet access.

Jerry's family frequented the library often, as everyone in the family loved to read. After hearing about the wild success of the MID docking system at school, Julia wrote to the head librarian and introduced this concept to her. The library seemed like an obvious place to implement it. Intrigued, the librarian contacted Julia within a few days. Julia told her about the success of the program at the middle school and high school, and offered to demonstrate the new technology to the library board and staff. It was a hit.

Within a year, the library had installed docking stations with a monitor, keyboard, and mouse at various locations throughout the building. People brought their MID, NetTop, or Pocket PC and plugged it into the docking station. Now it was possible to do searching, reserving, and requesting a hold from their own MID. They could use these MIDs to read e-books as well. In the children's library, the librarians came up with a fantastic application for school-aged children. Using the MID, they would hold a story hour using the paper book, while displaying each page of the e-book version of the same book on a screen, allowing all the children to follow along with each word as it was being read, and viewing the much larger pictures on the screen.

The library began holding a literacy night for adults using the same system. The program was a huge success. Since these MIDs have cameras installed, people could use this device to scan a barcode on a book as well. Similar to the supermarket, they could use their device for self-checkout. In addition, when they returned the books, they could scan all the barcodes on the books and then transmit this information to the librarian. After scanning, they would simply drop the books inside the drop box and then transfer their return list to the system in the library. With these new facilities, the return entries would be automatically updated in the system and show up as "just returned" when others searched for that book using the library system.

Julia's business was going so well these days that she decided to hire a manager to take over most of her duties, and take two years off to pursue her MBA. She had been eyeing the program at Stanford for a long time now, and, after discussing this extensively with Jerry, she decided to apply. Juggling a full-time MBA program with family life and keeping her business afloat would take a lot, but she worked out all the details and decided to go for it. Julia felt that she had a solid enough understanding of fashion design that she could effectively manage that part of her company, but when it came to expanding, opening new stores, accounting, and advertising, she felt that her skills weren't enough. She was also beginning to dream about the next step in her career and often fantasized about starting a consulting firm, helping others start up their dream businesses. She applied to Stanford's competitive MBA program and was admitted.

When fall rolled around, Julia went to the campus to register for classes and meet some of her professors. She was nervous, as Stanford was a large campus and she didn't know her way around, or have any clue how to sign up for courses. Similar to Jonathan's high school, Stanford had an application for the MID that included campus maps, names and contact info of all professors, and even the entire library catalog. Julia was thrilled. She realized she could even use this application to register for classes. All of the open classes were shown for her program, as well as the classes that were full. She could see an outline for her specific program that let her know what classes she should be taking and when. Julia registered for classes within ten minutes and then used the MID to locate the offices of some of her professors.

The MBA program comprised a mix of large lecture-format classes and smaller group sections. In the first month, Julia was assigned to work collaboratively with two other students in one of her smaller classes. The assignment was to address a current limitation that existed in the business world and to offer an appropriate business solution. It was very open ended. The professor stressed that the groups should come up with as innovative a method of presentation as they possibly could. In fact, the impetus behind the assignment was to get students thinking about new ways of presenting information. Later on, the professor hoped that students would incorporate each others' presentation ideas in future projects. As the program involved extensive practice in delivering presentations, he wanted his students to begin thinking outside the box in this realm. Each group would have to present their model to the small section in a week. The group that came up with the most interesting and applicable presentation would be asked to give the same presentation in front of the entire MBA student body and faculty.

When Julia's group got together to discuss the assignment, she told her group members about the MID docking station in her kids' classrooms. Julia had noticed that these stations did not yet exist at Stanford. Her fellow students were very curious about this, and decided to give their presentation using this system. The students had MIDs. Their classroom already had a large TV, computer keyboards, and a mouse. All they would need was the docking station. One of the collaborators had a connection at a tech manufacturer and was pretty sure that he could get one donated. The group decided that they would use this system to give their presentation, and then build their talk around how this same system could transform the way in which ideas were disseminated among colleagues at business meetings and in classrooms such as this.

Julia's group met three times that week, and in the end came up with an extremely sophisticated presentation that incorporated videos from the Web, real-time footage from other parts of the world, and Google Earth. On the day of the presentations, there were some fantastic ideas incorporating technology -- PowerPoint presentations and slideshows using laptops and a projector. Julia and her group had come in early to set up the docking station, and when it came their turn to present, they turned on the widescreen TV and explained what they were doing as they plugged Julia's MID into the docking station. They used the small MID to show the class how this new technology could indeed change classrooms and board meetings around the world. The rest of the class was entranced with the elegance of their presentation, and the ideas involved. The professor was impressed and let them know.

"This could become the new norm for delivering presentations. I'm going to talk to the dean about what you have shown us here, and work on getting these docking stations set up in our classrooms." He also chose Julia's group to give the same presentation in front of the entire MBA program.

SECURITY

It was spring break. Nancy and Jonathan were enjoying a week away from school. The family had decided not to plan a vacation, but rather just enjoy each other's company at home. Jerry and Julia had taken the week off as well. It was a gorgeous March day in the Bay Area, with temperatures in the mid-70s. The family was hanging around in the backyard, grilling fish and vegetables and playing badminton. It was an idyllic afternoon and everyone was relaxed.

As they were eating their dinner, Julia said, "Let's have a family check-in." Julia really loved having time like this with her family, and she so much appreciated hearing the news from her children and husband that sometimes didn't get told when times were busy. Jerry and the kids appreciated it, too.

Nancy, who was always eager to talk and share about herself, said: "I'll start!"

Everyone looked her way and eagerly awaited her news.

"A few days ago, one of the kids from our school was kidnapped," Nancy said.

Jerry, Julia, and Jonathan looked at her in complete surprise. This was not at all what they had been expecting to hear, and at first Jonathan thought she was kidding. Usually Nancy's stories revolved around the sports games her school had won recently, or a science project she'd been working on. Nancy continued with her story, and Jonathan realized Nancy wasn't kidding.

"This girl, Emma, was walking home after school on Tuesday. She lives on a street that's kind of deserted. Well, not many other people live on the street. The kidnapper drove up really close to her. Maybe he'd been following her, waiting until she turned onto an empty street. Anyway, he stopped the car, got out really fast, pushed her into the back seat and drove off."

"Oh my goodness," said Julia. She and Jerry exchanged glances.

They had, of course, heard stories like this before, but never so close to home. No one was eating their dinner. All eyes and ears were on Nancy, who continued telling the story.

Emma realized that she had been kidnapped. She's really smart, and pretty tough. According to her, the first thing she did was turn off the ringer on her MID and hide it under her shirt.

Then Emma asked the guy, "Why are you taking me away from my home?"

He said, "Stay down and shut up. I'm kidnapping you for money."

Emma told him that her parents had lots of money and that they really loved her, so she was sure they'd give him exactly what he wanted. She could see in the rearview mirror that he smiled when she said that. he believed her.

"Then," Nancy continued, "get this. Emma says she's hungry and wants some pizza. Can you believe she had the nerve to do that?"

"Wow," said Jerry. "She sounds pretty savvy."

"So the guy stops at a store to get her a pizza, always keeping his eye on the car to make sure she doesn't run away. While he's in there, Emma sends her family a message saying that she has been kidnapped and to look at the Locate Family application and find her right away. Then she puts the MID back on her shirt, making sure that the ringer is still off.

Emma's family has the Locate Family application, just like we do, so her MID was sending her location to the server every once in a while, just like it does for us. Emma's parents got the message, and immediately alerted the police, who came to get Emma's Mom in a patrol car. As they were driving, she kept looking at the Family Finder application on her MID. Using that device, the police surrounded the kidnapper's car and arrested him.

"And," continued Nancy, "get this...Emma thanked the guy for the pizza! Can you believe that? The kidnapper still has no clue how they found him."

"Wow," exclaimed Julia. "That's a pretty scary story. I hope that none of us ever have to use our MIDs in that way. And yet, what a relief it is for me that you kids could do the same as Emma if you needed to. That story is a great lesson for all of us. We should send Emma and her family a note with some cookies. That girl deserves it."

Now Jonathan said that he had a story to tell. Still rattled by Nancy's story, Jerry and Julia turned their attention to Jonathan, hoping that his story would be more pleasant.

"So this guy I used to know, Tim, dropped out of high school last year," Jonathan began.

"I remember him from when you guys were younger," said Jerry. "You were on the same soccer team for a while, weren't you?"

"Exactly," said Jonathan. "Anyway, he was working in that pizza place downtown. A few days back, he had an argument with the owner of the pizzeria and got fired. Supposedly, Tim was really angry and told his girlfriend Clara that he was going to kill somebody."

"The night after he got fired, Tim was hanging around outside of his apartment, talking to a few people and drinking beer. He picked a fight with one of the guys, then took a small revolver out of his jacket and started shooting randomly at trees and walls. But one bullet hit a guy on the arm and he started bleeding."

"Tim thought he had killed the guy. He got scared, hopped in his car, and took off."

At this point in the story, Nancy, Jerry, and Julia had totally stopped eating; their food was getting cold.

Jonathan continued, "Someone called 911. Since there was a man with a gun and he had opened fire, the police showed up with guns, and shouted at everyone to leave."

"Some of the cops started talking to people in his apartment building, which led them to Clara, Tim's girlfriend. She told them that Tim had been really upset about getting fired, but he had never planned to hurt anyone, and urged the police to find Tim as soon as they could."

"While speaking with the police, Clara had a moment of insight. She went into her apartment and brought out her MID, quickly launching the Locate My Friends application. Immediately, Tim's location popped up. He was driving north toward San Francisco. It only took a couple of hours for the police to track him down. They arrested him and admitted him into a mental Institute."

"Holy moley," said Nancy.

Julia said, "I remember getting a voice message from the city the other day, warning that there was a man with a gun in the vicinity, and advising residents to stay alert. Then there was a second message that he had been apprehended. I didn't want you guys to worry, so I didn't mention it."

Jerry added, "I wonder why the city police send the alert to people at home? It seems like they are probably the safest. It's those out driving and walking around who should get the alert. I mean, we didn't even get the message until the whole ordeal was over with. That's not too efficient."

Always interested in helping to make his local community work better, Jerry sent an email immediately after dinner to the city police, suggesting a new tactic. Alerting people connected to the city wireless infrastructure would mean that everyone in the vicinity using their MIDs would get the message immediately, and know how to keep themselves safe. The police force responded the very next day to his suggestion, and let him know they loved the idea, and would begin work on implementing that system as soon as possible.

"Well," said Jerry. "Since we seem to be on the topic of people breaking the law, I guess I'll share something I read recently on the Web. Don't worry, though. I really don't think it's as scary as your stories so far."

Julia looked relieved. All this talk of guns and kidnapping was making her feel slightly uneasy.

Jerry began, "There was a van heading from San Diego to Los Angeles. Wireless technology was registering that not only was the van speeding on the highway, but that it wasn't stopping at red lights. Now, we're all aware that these days, the police wireless monitoring equipment can issue tickets without even pulling the car over. Such was the case here; the van was issued several tickets within the course of an hour."

"Here's what I thought was really interesting. The Highway Patrol has an important component in their ticketing system. If the driver ignores the tickets and does not reduce his speed, or stop running red lights, an alert is immediately sent to the local police, and that is exactly what happened."

"Highway Patrol got an alert that this van was going at a very high speed, ignoring repeated violations, and running red lights. Using police cars and helicopters, the van was apprehended on the highway. After searching the vehicle, the police found that it was being used to smuggle goods into the country. The driver was immediately arrested and sent to prison."

"Well that's really a smart system," said Julia.

The kids agreed.

Julia thought for a moment and said, "Well, I'd like to stay on topic here. It would be out of place if I started telling you a story about a new employee we hired at the store. Oh, yes! I know what I can share."

Everyone settled in to listen.

"I was listening to "Talk of the Nation" earlier today on NPR while I was driving. They were discussing some interesting ideas about the prisons in California, and the amount of tax money that is required to house inmates. They brought up a fascinating solution that I hadn't heard about before. The experts mentioned that one system they are beginning to implement involves getting certain non-violent offenders out of the prison system, and yet keeping them under house arrest by using GPS technology to monitor their whereabouts at all times. The idea is that these offenders are then paying their own way, freeing up our tax dollars to be used elsewhere, and yet are still sort of in jail, in their own home."

Julia continued, "I thought of Mobile Internet Devices and the Locate Me applications. It seemed to me that this system would be better than GPS, as it would allow for two-way communication.

So I called the radio station and told them about my suggestion. They actually put my call on the air! Within a few hours, I got a call from the sheriff. He requested that I come to the San Francisco Police Department to demonstrate what I was talking about, so I've got an appointment to do that tomorrow."

"Wow, Mom!" said Jonathan. "That's really great!" Jerry and Nancy agreed.

The next day, Julia went to the sheriff's office and showed him her Mobile Internet Device. She explained: "Offenders under house arrest will be carrying this MID all the time. The device periodically sends date, time, and location information to the server. The history of the person's exact movements is automatically recorded in the server. At any time, the police officer in charge of the criminal can see it.

"Since, with these devices, two-way communication is possible, the sheriff can contact the person any time and talk to him. If the MID shows he is at home, the police can phone the home number to check on him. If the MID shows that he is at work (if he is allowed to work), the police can call him at work to check on him. The sheriff can also, at random, ask the local police to verify this person's presence at the location specified on the MID by driving there."

The police really appreciated Julia's idea, and promised her that they would begin implementing this new technology.

SHOPPING

When the weekend rolled around, the family piled into the minivan and headed out to do some shopping. Sometimes Nancy and Jonathan opted to stay home during these shopping escapades, but today Nancy needed a few school supplies and Jonathan wanted some materials to construct his latest science project. Julia and Nancy dropped Jonathan and Jerry off at Home Depot and went on to Costco, where they stocked up on groceries, bulk household items, and Nancy's school supplies. At Home Depot, Jerry and Jonathan were searching for the proper materials to fix the TV antenna to begin receiving digital programming, as well as Jonathan's items. It was a busy Saturday and the stores were crowded.

As they entered Costco, Julia instinctively reached for her wallet to show the employee stationed at the door her Costco card. When she pulled out her wallet, she laughed out loud at how thin it seemed, and remembered that she no longer had a Costco card. Instead, Costco had provided an application for her Mobile Internet Device, which she had already installed, so she could enter the store without showing her card.

There was now a scanning device at the entrance of the store. When the scanning device picked up the customer identification code in the MID application, it allowed the person to walk on through. However, if the scanner could not pick up an ID, a blinking red light went off and an employee came over. Julia and Nancy headed into the store without a problem.

"How cool," said Nancy. "It was always kind of annoying to have to pull out your card at the entrance."

"I agree," said Julia, "and I can't believe how light my wallet is!"

These days, the only thing Julia carried in her wallet was cash. Not long ago, it had been crammed full of stuff: her license, debit cards, credit cards, Costco membership card, library cards, coupons, phone numbers. Not long ago, on a trip like this, Julia would have a stack of Costco coupons that had been sent to her by mail. She always looked over these coupons carefully and saved the ones she thought were good, making sure to use them by the date printed at the bottom. She always kept the receipts from her previous shopping trips in her wallet as well, in case something needed to be returned. Now all that plastic was gone, the information transferred onto her MID.

Julia and Nancy loaded up their cart and headed to the check-out counter. Julia noticed immediately that the usual bustle of the checkout process seemed somehow more manageable this time than on her last trip. Usually the check-out lines were long, and customers were racing each other to get there first. This time there was no line, and things seemed to flow very smoothly. In the old days, Julia had to hand her Costco card to the check-out employee, and then fish around in her wallet for the proper coupons. Once all the items had been scanned, the employee would tell Julia how much she owed. Julia would pull out her debit card, slide it through the machine, enter her PIN, and then wait as the debit information was sent to the bank for authorization. Once approved by the bank, the check-out equipment printed up the bill and Julia put the receipt in a safe place. These days, Costco sent coupons directly to the customer's MIDs, saving paper, ink, and postage. On the MID, the coupons were well-organized and extremely easy to use. The selected coupons were automatically submitted to the check-out equipment, with the date and the specific items purchased.

This time, the check-out employee began scanning all the items in Julia's cart. Meanwhile, Julia took out her MID and held it up to the Point of Sale (POS) check-out equipment. Immediately, the MID established a connection. This meant that she no longer had to show the employee her membership card when checking out. Julia's coupons were automatically submitted to the POS equipment, and the machine accepted the coupons, immediately discounting the savings from Julia's total.

"Your total will be $127.59," said the employee.

Julia reached for her debit card to slide through the equipment. Pulling out her wallet, she was again surprised by its thinness and then remembered that she had just gotten rid of all her debit cards, and transferred all that data onto her MID. Julia laughed at herself again, and returned to her MID to complete the transaction. Using the MID, she launched the debit card application. Julia's information was immediately passed from her MID to the check-out equipment. The POS machine at the counter processed the money as usual, sending the card information to the bank to be authorized and approved. Once this was done, the bill was prepared. Even the final receipt was electronic. Julia saw it pop up on her MID and saved it in a special folder for receipts. The process had been quick, efficient, and smooth.

A major advantage to this new system was that all of Julia's credit card information stayed with her all the time. She never needed to get out a card to hand it to someone, which meant that her information remained safer. Even if someone managed to steal her MID, Julia made sure that all of her important information was only accessible by password, and she changed her password often. And, she was very good about making sure that her MID was always up-to-date, with appropriate firewalls, so that hacking into it was virtually impossible.

Julia especially loved that she could organize all of her receipts electronically. This meant that she never had to worry about storing them in a safe place at home (where quite often they got lost, no matter how hard she tried to keep them organized). In addition, she could download these receipts to the software she used to keep track of the family's finances. As a self-employed business owner, Julia needed to keep all of her receipts in order to do her taxes at the end of the year. Those complicated old days were gone. Now her receipts were stored in her MID, and could easily be moved into her home computers.

As they left the store, Julia observed, "These days my MID is loaded with such critical information. I'm realizing how important it is to back everything up and protect it."

"Yeah," replied Nancy, "I've heard that's important to do."

"Luckily," continued her mother, "our MIDs actually came with a complete system of backup and restore service that we signed up for. I don't know if you knew that our family's MIDs are periodically backed up, and that 24/7, our information is totally protected. If we ever lose an MID or need to get a new one, we can simply restore all the information from the old one with a few clicks. I think this all fits into the Software As a Service idea your Dad talks about."

"That's a smart system," said Nancy.

Meanwhile, Jerry and Jonathan were at Home Depot, trolling the aisles and searching for the proper aluminum pipes to fix the TV antenna so they could receive public digital television programs. Often they thought they had found the appropriate aisle, but after scouring the shelves, they still couldn't find what they were looking for. If they saw a free Home Depot employee and questioned him or her, the reply was usually, "It's right around the corner."

Just as father and son were getting their most frustrated, Jonathan realized that they had not needed to wander around.

"Dad!" exclaimed Jonathan, "We forgot about our MIDs!"

Jerry realized he was right. GPS had helped them locate Home Depot, but he didn't think it could locate what they needed once they were inside the store. He was so accustomed to the old way of shopping, he didn't imagine that Home Depot might have a wireless system. Lo and behold, when he got out his MID, a message popped up asking if he would like to use the Home Depot Local Positioning System (LPS).

Jerry said "Yes" and then typed-in the name of the materials that he was looking for. Immediately the MID offered him the exact location, floor, and directions to reach the product he was looking for. The MID also told him how many were available, and at what price. He also noticed that Home Depot offered information about the same product from other stores. This would come in handy if the item was out of stock here. Now, he could compare prices at five different local stores. Jerry was amazed. All the prices were similar, so he didn't worry about getting a better deal somewhere else. The Home Depot was only one floor, but Jerry noticed the MID also indicated "level 1" for the product. He realized that if the store did have more than one level, it would be really important that the device be able to tell which one the product was on.

Jerry and Jonathan found what they were looking for in no time. While Jerry was working with an employee to get the rod off the top rack, Jonathan began typing into his MID the materials he needed for his science project. Each time, the LPS technology was incredibly accurate. Jonathan and he could compare prices on the same products at nearby stores, Jonathan proudly reported to his Dad that by shopping at Home Depot, he had saved $46.33.

As they were leaving the store with their purchases, Jerry said, "Wow. Shopping will never be the same again!" Jonathan laughed in agreement.

Meanwhile, Julia and Nancy were pulling out of the Costco parking lot when Julia realized that she was low on gas. Julia and Jerry always filled up their gas tank at Costco. The only problem with the Costco gas station was that there was always a very long line, probably because the gas was cheaper there than anywhere else. For every transaction, each customer needed to slide his Costco membership card. Then the gas station equipment would verify each person's membership. Following this, the customer needed to slide his debit card, enter his PIN, and wait until the bank authorized their information. Only then could they go about filling up their tank. At the end of the purchase, the machine tells the customer to wait while it prints up a paper bill. It's no wonder that each transaction takes time, and that the lines are always long. This time, Julia noticed that the line was quite short. In fact, at two pumps, there was no line at all. Julia realized that it must be due to the new system using MIDs at the check-out.

"Wow," Julia said to Nancy, "let's get gas. I've never seen such a short line here."

"Mom," said Nancy, "there probably won't ever be a long line here again."

Nancy was probably right. Julia pulled up to the gas pump, turned off the ignition, and popped open the gas tank. She took her MID close to the equipment; immediately, the MID established a connection. Within seconds, her membership was verified. She selected her debit card and the gas station sent her information to the bank for authorization. Then she was free to fill up her tank, after which the gas station equipment sent her MID an electronic receipt.

With Web on-the-go, the gas transaction was quick and easy. There was no need to slide a card or enter a PIN. There was no bill to print and keep track of. Julia and all the other customers were delighted with this new way of doing things. Over time, Costco shut down two of the pumps, saving money on equipment. Even then, transactions went so fast that there were no lines.

Julia and Nancy picked Jerry and Jonathan up at Home Depot. They found each other easily using the Family Finder application and shared success stories from their shopping trip so far. There was just one more errand to run, and it was on the way home. Jerry and Julia had to deposit some checks and get enough cash for the week.

Because they were no longer carrying their ATM cards around, the banking process was different as well. In the old days, Jerry and Julia would insert their debit card, enter the PIN, select the proper transaction, then deposit their checks or withdraw cash. These days, however, things have speeded up. When they got to the bank, Jerry hopped out of the car with the checks and his MID. As he approached the ATM, his device established a connection with the machine, identifying him as a valid customer. Jerry completed the entire transaction using his MID. He withdrew two hundred dollars and then deposited his checks into the ATM. He checked his balance on his MID, then closed the application and got back in the car.

"Wow," said Julia, "that was quick!"

When the Internet was launched, it was used mainly for sending and receiving emails. Then Instant Messenger and discussion forums became popular. Once people gained confidence with the Internet, they began using it for business applications. Banks launched online banking. Pizza stores launched online ordering. People used the Web to locate products and local stores. UPS and FedEx used the Internet to track items, and a whole array of online stores sprouted up.

The fact is, once the proper infrastructure was established and made available, reliable, quick, safe and secure, businesses began to take it seriously and began exploring new ways to make the Internet applicable to their specific enterprise. And people continue to come up with new ways of using the Internet to do business.

Similarly, once the Web on-the-go infrastructure is established, businesses will come up with their own ideas about how best to use this technology to make their specific enterprises run more efficiently.

BUSINESS ENTERPRISES

In this story, we'll take a look at how Web on-the-go changes business enterprises.

Most business enterprises enforce a tight system of security on their campuses. Each employee is required to wear a security badge at all times on the company premises. Without this security badge, an employee is barred from entering any of the buildings. So, along with all the other necessary items that one must carry—driver's license, bank debit and credit cards, cell phones, and car keys, employees must remember to bring their security badge to work every day.

Like all the other employees, Jerry sometimes forgot his badge. Because it didn't fit inside his wallet, it was easy to forget, and on an average of once each month, he found himself heading over to the receptionist's desk to ask for a temporary badge for the day. These badges were valid for only one day and then had to be returned. Because this happened to Jerry with such frequency, he got to know the receptionist pretty well. One day when he had forgotten his badge, he asked the receptionist if he was the only one in the company with such a bad memory.

She laughed and said: "Oh, no. There have been five people today who forgot their badge, and it's not even 9:00. I'd say on average, about twenty-five employees come asking for a temporary badge every day."

The secretary continued "And you should see it here after a three-day weekend. So many people come looking for a temporary badge that the line stretches out the door!"

Though this made Jerry feel better, it also got him thinking. His company had ten buildings, and this was just about one. He estimated that there were 250 people who forgot their badges every day.

These days, the long lines at the receptionist's desk are no longer an issue. Because a security badge is a plastic card containing specific data, similar to an ATM card, all of the information held within the badge has now been transferred to Jerry's MID in an application specific to Jerry's work. All employee security is now within this application. Now, right after Jerry parks his car in the employee parking lot, he gets out his MID and launches the company security application. As he approaches the front door of the building, he passes through scanning equipment that searches for his employee information on the MID. Upon finding Jerry's information and authenticating it, the door is automatically unlocked. If this information is not found, the person is asked to see the receptionist.

Using this application, business enterprises have gained a much higher level of security. All employees at Jerry's company have certain buildings and wings they are authorized to enter. For example, only the R&D engineers need access to their department, while senior-level executives only need access to their offices and certain meeting rooms. In each case, complete employee information is necessary in order to authorize a certain person to access a specific wing. Certain employees are authorized to enter multiple locations, while others are only authorized to enter one specific location.

Because the MID contains all the employee verification information, the scanning and security equipment is able to make the decision as to whether a certain employee is authorized to enter a specific wing of a building. For example, if an R&D engineer attempts to enter the wing where he or she is authorized to go, then immediately upon scanning the MID, the door will open. If the scanner does not pick up authorized access for the person trying to enter, the security equipment doesn't open the door. Not only that, but the equipment makes an entry in an electronic log about which employee attempted to access which location, and at what time and date. Similarly, if a meeting of only top-level executives is underway, the scanning security equipment will only allow entry to those employees who have proper access.

And, of course, there is no longer an issue with people forgetting their badges. The receptionists are freed up to focus on more important work, and the employees are thrilled that they no longer have to worry about forgetting their badges. Jerry couldn't be happier, to have all of his important documents stored electronically in his MID.

Jerry was having an extremely busy morning at work. Suddenly he got an alert that his presence was requested at a meeting happening in two hours in the main building. The meeting had been organized at the very last minute, and the notice said that it was extremely important that he attend. Jerry was clueless as to why the meeting had been called, which added an additional level of stress to his already stressful day. He finished up what he was working on, and, feeling rather tense, found himself a few minutes behind schedule. He would have to rush to make the meeting on time, so he opted to drive his car to the main building, instead of walking over or taking the shuttle. He parked in the lot closest to the main building and ran inside.

Jerry slipped into the meeting just as it was beginning. The top-level executives from his firm began giving a detailed presentation about the company's current expansion in Japan and other Far Eastern countries. Jerry had heard rumors that this was happening, but he hadn't been involved in any of the negotiations. When they were finished presenting the expansion plans, one of the executives addressed everyone present.

"You all have been chosen to help build the infrastructure in the new offices in Japan. You will be involved in all tasks, including human resources, training, team-building, and oversight of new employees, among other duties."

The executive went around the room, giving people specific jobs, locations, and timeframes for completion. Jerry was told that he would be asked to travel to Tokyo in order to help train employees. He would be there for two weeks, followed by a week or two in other countries nearby. Jerry was totally shocked and surprised. He had always wanted to visit those places. He even enjoyed eating sushi. A fully paid visit to Japan was a dream come true. As the meeting let out, he lingered behind and thanked the executives for the opportunity to be a part of the expansion of the company. He could not help but let his enthusiasm show.

"I'll do everything I can to build an effective, result-oriented team in Japan," he promised.

Reeling with the excitement of this news, Jerry left the building. He couldn't wait to tell Julia and the kids. The shuttle was picking up employees to take them to other buildings on the campus. He got on with a few other people, dreaming of his trip. He got off at his building and nearly flew to his cubicle, where he immediately emailed his friends and family about the news. He could hardly believe that in a few months he would be traveling around Japan, South Korea, Hong Kong, and Singapore. He wanted to do a bit of sightseeing and was already planning how to do that.

Throughout the day, Jerry couldn't stop thinking about the trip. By six o'clock, he was tired and even though he had a lot left to do, he decided to call it quits for the day. He left the building to go to his car, but it was not in its usual spot. He walked all the way around the building, couldn't find it anywhere, and was beginning to think that it had been stolen. He called Julia.

"Julia," he said, "I can't find my car! I think it might have been stolen!"

"Really?" exclaimed Julia. "That seems unlikely. It's pretty secure at your building. Why don't you try using the MID to find it?"

"Of course!" said Jerry, "I didn't remember that. I'm not really myself right now with all this excitement."

Jerry launched the Locate My Car application on his MID. It showed that his car was near the main building. Jerry had rushed to get to the meeting and had driven over there, and then, excited by all the news, had mistakenly taken the shuttle back. He couldn't believe he had forgotten. He forgave himself, however, realizing that the day had been unique.

Still on the phone with Julia, he said, "Oh, my. I can't believe I didn't remember that! I drove over to the main building earlier today."

After hanging up, he walked over to the main building. Sure enough, his car was the only one left in the parking lot. The MID Locate My Car application is very useful for people who tend to forget things in the midst of their busy lives. Since that incident, Jerry always uses the Locate My Car application when he finds himself in a similar situation.

Later that week, Jerry got a call from Linda, the director of the company's Travel Department.

"Your passport needs to be renewed this year," she said to him. "Why don't you go ahead and apply for the new electronic passport?" She gave him instructions.

The very next day, Jerry went to the local Mountain View Post Office and filled out the application for the new electronic passport. The passport clerk entered Jerry's information into the computer, then took his picture using a digital camera and quickly scanned the supporting documents, including his birth certificate, into the computer.

Curious about the new system, Jerry asked, "So how does this work? I've only ever renewed my passport the old way."

"Well, it's similar to creating a bank account," the employee said. A bank clerk is authorized to create accounts. Once an account is created, an individual will have the secure authentication information to access their bank account. In addition, one is able to access one's bank account over the Web. The new passport accounts are exactly the same.

"So right now I'm creating a new account under your name in the United States Passport Server. I have been authorized, as a passport clerk, to create new accounts. I've created yours by entering all of your basic information, including first, last, and middle name, date of birth, home address, contact info, and so forth. And then I uploaded your photos and birth certificate."

"And then what?" asked Jerry.

"Then this new passport account needs to be approved by the Regional Director of the Passport Office. Once I'm done creating the new account, a workflow will be initiated to send a notification to the Regional Director, along with a message stating that he or she needs to take a look at the new account creation. Then they will either approve or reject the passport renewal, or ask for more information. Once the Regional Director approves your documents, your passport account will be created on the United States Passport Server. Authentication information will be sent to you. Using that information, you can install your electronic passport on any of your Mobile Internet Devices."

"Wow," replied Jerry. "So there will be no booklet?"

"Nope," replied the clerk, "no paper whatsoever."

Jerry was really excited and enthusiastic about this new technology, but he found himself wondering what would happen in the other countries that he would be visiting. Would they have trouble accessing his passport?

Jerry decided that he would ask Linda these questions, as there was a line behind him waiting to talk to the clerk.

Before he left the Post Office, Jerry got an email notification on his MID saying that a new account under his name had been created in the United States Passport Server and was currently waiting for approval. The next morning, Jerry received another notification saying that the passport application had been approved and that his account had been created. In the same notification was the authentication information. Using that information, Jerry installed his electronic passport on his Mobile Internet Device.

Jerry was truly amazed. In the past, renewing a passport was a very lengthy process requiring much planning ahead. And if one wanted the process expedited, it was very expensive. This had taken only 24 hours!

In the afternoon, Jerry had a meeting with Linda in his company's Travel Department. He expressed his concern and doubt regarding the feasibility of the electronic passport when traveling overseas. Linda soothed his worries.

"This is the latest technology that has been introduced to Homeland Security and the State Department. Because many Silicon Valley companies were crucial to creating this technology, all the companies in the Bay Area decided to promote this electronic passport. This technology needs to start somewhere, and we decided, 'Why not here?' So in all honesty, that's the reason I suggested getting this new electronic passport."

"Don't worry," she continued. "I recently visited Japan and some of the other countries that you are going to. Look, I'll show you."

Linda took out her MID and showed Jerry her electronic passport.

"So first, at the San Francisco Airport, when I was leaving for Japan, I showed the electronic passport in my MID to the authorities there. They received the authentication information without a hitch, and using that information, pulled up my record from the United States Passport Server. Then they verified my passport. Now everything is global. Just like global warming, we have a global economy, global finance, global travel, and global security. To meet the challenges of global travel, all the countries have agreed to set up and maintain the Global Traveling System. All the airport authorities around the world have access to this system, called GTS," Linda explained.

"GTS helps to track all global travelers. It captures the date of departure from one's home country and the date of arrival at the country being visited. All the airport authorities have access to this GTS, and can view the traveling history of any traveler."

But Jerry didn't like the idea of some system tracking his entire traveling history, and he let Linda know that.

"You're not alone. No one likes this global tracking idea. But it helps the authorities to monitor international security. It is just like your credit history. The credit score system keeps track of your credit history and makes it available to the authorities. The GTS keeps track of your traveling record and maintains it. Thus authorities from any country can see when you do or don't misuse your visa, when you arrive on time, and leave the country on time, and whether you have a record of smuggling illegal goods.

"So, after verifying my electronic passport, the San Francisco Airport authorities accessed the GTS and made an entry that I was leaving the U.S., including the date, time, flight number, and destination country. They also asked for my expected date of return. All this information was recorded in GTS."

"Then when I landed in Japan," Linda continued, "I was surprised to see that the Japan Airport authorities also had the ability to access my complete information from the United States Passport Server. I asked about this, because it was concerning to me that they had access to all my personal info. They told me they didn't have the ability to manipulate data on a person from another country, but they did have read-only access. Using my electronic passport, they pulled my record and verified. Then the Japan Airport authorities accessed my record in the GTS and updated it with my Tokyo arrival date, time, and information."

"You will learn more about the GTS during your travels. One major advantage is that no one needs to enter any data manually into any system. They just scan your MID and information is transferred. That's the main reason for the success of the GTS. If it had been a manual entry system, then it probably would have failed. No airport authorities have the patience to enter all that information into the GTS."

Linda realized that she had just talked Jerry's ear off with a lot of technical details, so she decided to change the subject.

"It's interesting to think about how human beings are the only living things that carry these passports around with them when traveling. I mean, look at birds. Look at whales! They just move around the Earth following their instincts. They have complete freedom."

"Think about it: the words world, country, citizen, borders, and even global citizen are all political concepts created by men that divide people. Birds and animals approach this planet as one whole Earth. Men divided the Earth into countries, but Planet Earth unites the human beings. If aliens from other planets took a look at us, they would just see us as human beings from Planet Earth, not as citizens of one country or another."

"Maybe when every human being realizes this truth, then we won't need all this sophisticated GTS technology and electronic passports. Until then, we're doing our best. Good luck to you! Enjoy your trip!"

"Thank you," said Jerry. "You've been very helpful. I like what you said, especially at the end."

Linda replied, "This passport stuff is what I do for a living, but that other stuff is how I try to live my life."

Airlines used to give passengers printed tickets, sent through the mail. A traveler had to hold on tight to his ticket and bring it the day of his flight. After that came the electronic ticket. Using the electronic ticket system, an airline or travel agency sent ticket information and itinerary to the traveler via email. The traveler needed to print out the emailed itinerary and make sure they brought it to the check-in counter. Though it was called an electronic ticket, it still required printing something. In addition, the traveler was required to provide photo identification at the ticket counter and also when passing through security. It was possible to use a driver's license or passport for this photo requirement.

These days, things have changed. The day after his conversation with Linda, Jerry got an email from the Travel Department at his company regarding his airline ticket and visa information. Since it was the first time Jerry had traveled abroad with his company, the Travel Department asked Jerry to come to their office to go over everything. There, Monica, the travel assistant, installed a Travel Application on Jerry's MID and linked it to the electronic passport. Then Monica transferred all the proper visa information for the countries Jerry was going to visit, including Japan, South Korea, Hong

Kong, and Singapore. Finally, she transferred the airline ticket information to his MID as well.

All Jerry needed to do was show up at the airport with his MID, which was now fully loaded with all of his traveling information. He could even check in the night before his flight from his MID. And, he learned, if there was a change about the flight, he would get an alert on his MID. Monica told him that it didn't hurt to check the flight online as well.

Finally, the day of Jerry's departure had arrived. Linda had walked him through the entire process of traveling with an electronic passport step by step, so there weren't any surprises as he made his way through the airports. Using his electronic passport and electronic ticket, he boarded the airplane to Japan. It was a ten-hour journey. On the flight, there was a movie, video games, and other entertainment. Not only that, but there was also wireless Internet access. Jerry found this extremely helpful, as he had some last-minute work to do in preparation for his upcoming training in Japan. He was also comforted by the fact that he could keep in touch so easily with Julia and the kids. He checked the news quite often and watched some YouTube videos.

After ten hours, he touched down in Tokyo. Jerry flew through Customs and Immigration. He had never had such a smooth overseas traveling experience. And, the information for this trip was now recorded in the GTS. He knew the name of his hotel, and typed the hotel name into his MID. Immediately up popped the address and directions. In some countries he had traveled to, Jerry had noticed that the local cab drivers recognized him as a tourist and treated him accordingly.

Once, a cab driver had taken him on an hour-long drive before stopping at the hotel, charging him for the entire ride. Later he found out that the hotel was only ten minutes from the airport! He had even been able to see the airport from the balcony of his hotel room. This time, Jerry felt more confident. He knew not only the address, but also the mileage and driving time, and thus knew he couldn't get jerked around.

While on the flight, Jerry had researched all the hotels that he wanted to stay at in Tokyo and beyond. He had also installed an application called the Japanese Translator. This handy little application was a text-to-translate-speech tool. Jerry had already typed a few regular expressions like, "Hello, I want to go to this hotel. Thanks." When Jerry selected an expression, his

MID would translate it and immediately convert it into a voice message in Japanese and then say it out loud.

When Jerry selected "Hello," the speaker said "Konichiva."

Using this translator, Jerry easily found a cab and reached his hotel. He ate a fantastic dinner and got a good night's sleep. Jerry's first week in the Japanese office had been full of fun and hard work. Now it was Saturday, his first weekend in Japan. He decided he would explore the city on his own, and had already discovered that Tokyo provided a service called City Guide for MID. Using his MID, Jerry connected to City Guide and began his tour of the city.

Based on the GPS location of Tokyo, the MID accessed information and the history about that specific landmark. Jerry could even select a picture from a photo gallery depicting how that same location would have looked forty years ago. Jerry had a blast exploring Tokyo on his own. The City Guide provided all the information he could possibly want about restaurants as well.

It was like a real person walking along with him, providing all the information that was interesting about the city.

After spending a few weeks in Japan, Jerry spent a few very exciting days in South Korea, a couple of marvelous nights in Hong Kong, and finally arrived in Singapore, the final stop in his multi-city business trip.

At the Singapore airport, an Immigration authority told him: "Singapore has the best hospitality in the world, if you are well-behaved. But Singapore is not afraid to punish you if you break the law. I see here that you are planning to stay here in Singapore for two weeks. You need to leave the country before the end of two weeks. Recently a tourist drank too much and overslept in his hotel room. His visa expired by only a few hours, but still he was arrested and received the caning punishment."

Jerry was taken aback, and decided that would definitely not happen to him.

Then the authorities installed an application on Jerry's MID called Tracking Visitors. They explained that Jerry needed to carry his MID with him all the time, everywhere he went, whether to work or to a restaurant. On a daily basis, this application would capture his GPS location and send it to the server. Thus, if the authorities needed to locate any tourists for any reason, they could do so.

Jerry did not like this idea and argued a little bit. "Well, that's not really what I would call freedom," he said.

The authority responded, "The good citizens of Singapore enjoy total freedom. But we cannot give the same kind of freedom to all of our visitors. We didn't have these same rules twenty-five

years ago. Right now, however, the world is changing, and harmful people are finding ways to travel the world. We need to enforce this technolpgy for the security of our country."

Jerry saw the truth in this and so dropped his argument right there. He was not interested in getting into a dispute, especially after hearing about the tourist who overslept.

"After a week," continued the employee, "this application will remind you that you need to leave by the following week. If you overstay your time, after your visa has expired, this application will send us your GPS location every hour."

Jerry always respected the laws of whatever country he was in. Wherever he went, he took special care to understand the local laws and respect it. In the end, he had a very pleasant stay in Singapore and a lot of fun.

On his way back from Singapore to San Francisco, he prepared an email and sent it to the Homeland Security Department. He had an idea. He told them that he had recently become aware of a new method of tracking visitors to other countries. He brought up the recent concerns in America regarding immigrants from Mexico legally entering into the U.S. with a work permit but then failing to return before the visa expired, thus remaining in the U.S. as illegal immigrants. Just like in Singapore, Jerry offered, the Homeland Security authorities could give an MID with a similar application installed to these immigrants. It would be like an electronic work permit. That application would send the daily GPS location of that immigrant to the server. Using this technology, the authorities could easily track them. Once the visa expired, the MID would send alerts on hourly basis. Thus the comings and goings of illegal immigrants with work permits could be controlled fairly easily.

The Homeland Security authorities really liked Jerry's suggestion and wrote back in a few days to let him know that they planned to research and implement it as soon as possible.

CITY UTILITIES

In this story, we'll take a look at how Web on-the-go could change City Utilities.

It was a beautiful Saturday morning. There wasn't a cloud in the sky and a slight breeze kept the temperatures in the mid-70s. Jonathan was playing soccer with his high school team at the city park soccer field. Soccer was Jonathan's sport of choice, and he excelled at it. The entire family had gathered to watch the game. Julia was sitting next to another soccer mom, Lisa. Lisa and Julia had children of similar ages and had known each other for many years. They enjoyed each other's company and made sure to sit next to each other at all the games.

"I'm not sure that I ever told you this," Julia said to Lisa, "but my Dad was a professional soccer player."

"Wow!" said Lisa, "That's really cool!"

"When I was a little kid, I used to go watch him play his big games. It was so exciting. Whenever I come to one of Jonathan's games, I think about him. I know he and my Mom would really love to be here to watch him play, but they're in the UK! We always email them immediately afterwards with details."

"Actually," said Lisa, "they could see pictures from the game right now, even from the UK. Look at those posts on the side of the field. Right now there are four digital cameras taking pictures of this game, and there is another piece of equipment collecting the pictures from the digital cameras and uploading them onto the city Website. Since the park has a wireless connection, all these things are possible now. The players and family are able to share the pictures with their family and friends."

Immediately, Julia took out her MID and launched the browser in order to view the city's Website. She was thrilled to see pictures of Jonathan playing soccer in the field where they were now sitting. She saw pictures even from a few moments ago. Coincidentally, Julia noticed that her father was online and available via Instant Messenger. Julia immediately forwarded him the link, and shared the pictures with him. Her dad wrote back in a few moments, extremely enthusiastic, and happy to have been included.

"If they can do this," said Julia, "I don't see why they couldn't install video cameras as well."

"I've actually heard rumors that they are working on that right now," said Lisa.

Within a few months, the city had done just that. The cameras on the soccer field captured the games being played on video and uploaded it into their equipment. This meant that Jonathan's grandparents living on the other side of the world could watch a streaming video of him playing soccer, almost in real time.

Though Jerry doesn't have a lengthy commute, he does find himself driving to and from work pretty much every day. Because he drives the same route each day, he has memorized all the stoplights. As he leaves his neighborhood, Jerry knows that there won't be much traffic to speak of, as the streets around his house are not that populated. However, he also knows that he must stop at every single red light. He had once received a ticket around here for slowing down, but not thoroughly stopping, at a signal, so these days he is extremely careful, and always makes sure to come to a complete stop.

It used to be that Jerry caught all the red lights, even if there wasn't another car or pedestrian in sight, and this really annoyed him. It made his commute time much longer. Jerry knew how the system worked, because he'd asked the cop who pulled him over.

When Jerry's car approached the street crossing and stopped at the red light, he actually stopped the car on top of a sensor in the road. This sensor would send a message to the traffic signal to change to green, and then he was allowed to keep driving. In the past few months, Jerry had seen utility trucks out and about, doing a lot of work with the traffic signals. And then,

seemingly overnight, he noticed that all the traffic lights on his commute were green. There was no more waiting needlessly at red lights. His commute time was drastically reduced. Always curious, and interested in how things worked, Jerry checked in with the city government officials to see what had changed. The city official on the phone was very patient, and also proud to be explaining the recent upgrades.

"These days," the woman explained, "we have a wireless traffic control system. As you know," continued the city official, "the old sensors on the road could only alert the traffic signal when the weight of the car was on top of it. These days, when your car approaches a traffic signal, the traffic control system checks for other cars approaching from other directions. If all the other streets are free of cars, then the light automatically turns green."

"Basically," the woman continued, "the wireless traffic control system is able to receive the information from your car well before your car actually arrives at the light and is able to process all of this information well before you would even begin slowing down for a red light. This allows the driver to keep driving at a smooth and steady pace, without making unnecessary stops along the way. Not only that, but it saves drivers gasoline as well."

"Well," asked Jerry, "that's fantastic. What about when there are numerous cars in one direction?"

"Let's assume that there are ten cars in one direction," replied the woman. "With our old system, the sensors in the road had no way to calculate whether or not there were more cars behind the current one. The sensor didn't know how many cars wanted to move forward with that same green light. Drivers who knew the system would drive forward really quickly in order to reach the sensor before the light changed. With the new, wireless traffic control system, the device is able to calculate the exact number of cars waiting in any direction and can then make the right decision. The drivers don't need to rush forward in order to keep pressure on the road sensor. If there are numerous cars in every direction, then the light will work on a timer, allowing each direction to move forward for a certain amount of time before switching."

"Wow," said Jerry. "I'm impressed."

"One more thing," the woman offered, "Since we are done using the buried sensor technology, we don't need to have city workers out there digging up the road every so often."

"Yes," Jerry replied, "that's a great added benefit."

It was a Wednesday evening. Julia drove to Nancy's school to pick her up from practice. Recently Nancy had been selected to play on the school's badminton team. Her gym teacher had noticed that she was really good at the game and had asked her if she would like to play on the team. Nancy was thrilled. She stayed after school three days a week to practice. While Nancy finished up, Julia had to wait in the car. She had arrived a bit early, and decided to take a walk along the small creek that ran alongside the school. While walking, she saw some birds that she had never seen before, fishing in the current, and got wrapped up in watching them. She took note of their colors and shapes so that she could look them up later.

After a while, Nancy came to the parking lot and looked for her Mom. She didn't find her, but did see the car. Nancy took out her MID and launched the Family Finder application. Julia was walking along the creek. Julia was pulled out of her bird-watching by an incoming message from Nancy.

"I'm done with practice and waiting near the car," read the text.

Nancy saw that Julia was quite a distance away from the car still, though coming closer, so Nancy knew she had gotten the message. Nancy was tired and wanted to sit inside the car till her Mom arrived. She wrote her Mom another note: "Can you open the car?"

Julia knew that the remote in her car key would work only if she pressed the key within ten feet of the vehicle. However, she remembered that Jerry had been talking the other day about being able to unlock the car using the MID. Using her MID, she accessed the lock system in the car, and selected "unlock."

"Try it now," she wrote to Nancy.

Julia knew that it had worked, as she could see a small icon of the car, and watched as the passenger side door on the icon opened and shut.

"This is the real remote," Julia thought to herself with amazement.

"I could lock or unlock the car from anywhere."

In the real world, OnStar, from General Motors, currently offers this type of service, the ability to unlock a car remotely. In case of an accident, they are able to inform the police immediately.

On these GM cars, a module installed in the car communicates with the OnStar system via the cellular network. However, as we have seen, this system only works when a car is inside the boundaries of the cellular network. In remote areas, this system does not function. In our Web on-the-go story, let's assume that every car has been outfitted with these OnStar service features, and that wireless technology allows the system to work just about everywhere.

In the car, Nancy turned on the radio and began listening to her favorite radio station. She could see that her Mom was getting closer, but was still a ways away. Since she had lots of free time, she started searching around in the car. She was curious about the insurance and registration documents that used to be in the glove box. She remembered they had been pulled over once and that these documents were out-of-date. When she looked around for the documents, she couldn't find them.

Finally Julia returned to the car.

"So where are all the insurance documents?" Nancy asked right away.

"Ahhh," replied Julia, "I see you've been doing a little snooping around."

"Well, now that the car has an embedded Mobile Internet Device, it is as good as a computer. All the documents, including the service documents, registration, and insurance documents are all stored in the car's MID. It even knows when the car needs to be serviced, and sends me reminders telling me when I need to take my car in."

"So you never let them get out-of-date!" said Nancy.

"Exactly," replied Julia and with that, they headed for home.

In order to get to work, Jerry had to cross railroad tracks. This railroad crossing had the potential to be very confusing. Sometimes, even though the traffic signal was green, there was a train approaching, which meant that the railroad warning system rang a bell with red lights on it, and a gate came down, barring anyone from crossing the tracks. Of course, in this instance, right of way was given to the train, and all vehicles had to stop and wait for it to pass, and for the gate to be lifted.

Today, as Jerry was driving to work, he approached the track just as the warning lights had begun to alert drivers of an approaching train. Because

Jerry always thought about the systems he encountered, he thought about the train crossing system as he was patiently waiting for all the boxcars to pass. He realized that the signal worked as a one-way communication system. The railway signal near the road received an alert that a train was approaching, and then immediately closed the gate. Jerry realized, however, that the train conductor had probably never received confirmation that the gate had been closed, and that it was safe to proceed. The conductor simply moved forward on the assumption that the gate had been closed. Luckily, for the most part, this system had worked out, and there hadn't been many accidents.

In fact, most of the time, the train drivers operate as if they are blindfolded, totally relying on the signals along the tracks. They proceed if they see a green light and they stop if they see a red light. The rest of the time, they can just close their eyes. They have no real information about what is happening in front of them, what is in their way, or whether the crossing is really clear.

However, thought Jerry, if I were driving the train, I'd want to get confirmation that the crossing had been closed and that it was indeed safe to go on. I'll bet that's possible with the new Web on-the-go technology. Jerry was getting excited now. He thought about how wireless equipment could be installed at the railroad crossing point and could send this confirmation to the train driver as soon as the crossing was closed. In fact, they could even install a digital camera. The train driver could have a TV-like control panel where he could also receive visual confirmation that the railroad crossing had closed and that it was safe to proceed.

Jerry's mind continued to work through this. He had traveled quite a bit on his own, and knew that in most developing countries, the railroad crossing gate was manual, or else there was no gate at all. If these countries could also have access to this technology, it would be much easier to have an automatic gate at the crossing. If the conductors had access to this type of confirmation information, Jerry was positive that the safety and security of the world's railroad system would improve dramatically.

Instead of working on blind faith, digital cameras could be installed every so often along the track. The digital cameras could send the real-time view of the track to the train drivers on the TV-like control panel. The train drivers would then have a much clearer vision of the crossing they were about to pass through. After all this thinking, Jerry noticed that the train had now passed. The gate went up, the light turned green, and Jerry continued on to work.

Having grown up in a very athletic family, Julia kept herself fit. A few years ago, she participated in the San Francisco Marathon. She trained for a long time and found the entire experience to be invigorating and fun. When she had shown up at the registration for the race, they had given her a chip that needed to be tied to her shoe.

Jerry had been with her and explained a bit of the history around the chip. Julia hadn't asked, and she didn't necessarily want all the details Jerry had to offer, but she knew that an important part of being in a relationship was letting the other person talk about the things that are interesting to them. She also enjoyed Jerry's ideas and knowledge.

"I've heard about this at work," Jerry began. "The basis for the marathon chip is called the radio frequency identification system (RFID). This is the exact same technology that is used for security-locks in cars and admission control in buildings. So the chip they just gave you is a miniature transponder. It contains a chip and an energizing coil. These two are encased in a waterproof glass capsule, which allows the marathon chip to be used in all kinds of weather conditions. I'm pretty sure that you can wear the chip in a bunch of ways. In the marathon, as you know, you attach it to your shoe, but if you wanted to participate in a triathlon, you'd probably have to wear it on an ankle bracelet."

"I think it's pretty cool that the chip you are wearing doesn't have any batteries. The transponder is passive until it moves into a magnetic field, generated by a send antenna. At that point the energizing coil in the chip you are wearing produces an electric current and this powers the chip. The transponder in turn sends its unique identification number to a receiving antenna. The whole procedure takes approximately 60 milliseconds and is repeated continuously."

Julia actually was interested and Jerry continued. He knew he had a receptive audience.

"The send and receive antennas are placed in thin tartan mats. These antenna mats are strategically placed at different points along the course, including the finish line. They are connected to a box at the side of the road that contains electronics and batteries. Each time an athlete wearing a marathon chip crosses one of these mats, the chip is energized and sends out its ID number. This number and corresponding time are then stored in the box, and immediately sent to a computer that is collecting all the data from all the participants."

"Really, what that little chip can measure is the time it will take you to finish the race."

Julia thought all of that was very cool. But she actually wanted to have more information about her running, such as her speed and pace, so that she could improve in the future. Julia did great in the marathon that year, placing in the top quarter of all the women who participated. This year, Julia decided she wanted to run the marathon again.

She looked into the application form, and saw that instead of using the old RFID chip, this time the marathon committee had decided to use Web on-the-go technology. Participants from the general public simply installed the San Francisco Marathon application on their MID and then wore their MID on their waist belts as they ran. They gave each professional runner like Julia a marathon chip-like device. In reality, this chip was a tiny MID with limited programming and abilities. It had no display and no keyboard. It was very simple, which allowed it to be small.

At the registration table, a marathon employee programmed Julia's ID code into the new marathon device. They also told the device how often the information should be sent to the Website, and the Web address where the information should be sent. Then it was explained how the new system worked. She could wear this new marathon device anywhere, on her shirt, shorts, shoe, cap, neck, or arm. Periodically (in her case, every thirty minutes, and for serious professionals, every five minutes), this device sends her identification code, current GPS location, and the time, to the server.

Jerry and the kids saw her start the race and then stationed themselves at a street where they knew the marathon would pass. Thanks to Web on-the-go technology and infrastructure, the city broadcast the whole event in real time, so Julia's family could watch the live television broadcast. Once they knew the event had begun, they began to track Julia's progress.

Since the new marathon device on Julia's waist belt periodically sent her location information to the Website, Jerry and the kids could watch Julia's movement on the city map. It also gave the graphical representation of her data. Jerry could see at what speed she started, where she picked up speed, and where she slowed down. It was even possible to compare her information and progress with the other participants in the race.

In the UK, Julia's parents were also watching Julia run in the San Francisco Marathon. They were delighted to be included in this major event, even from so far away.

Julia loved to get up early, make her coffee, and read the paper before the rest of her family got out of bed. One morning she found an article that she thought was really interesting. It was Sunday, so everyone was home, and sleeping in. When they finally woke up, they all made breakfast and gathered around the kitchen table. Julia could hardly wait to share with them the article she'd read.

"I saw something interesting in the paper this morning," she began.

"What was it?" asked Jonathan, still a bit sleepy.

"So we all know that up until this point, there has been a city employee who goes around to each home every month to get an updated power, water, and gas usage meter reading. Based on these readings, each household gets a bill every month. We see the PG&E folks come and do that all the time."

"But this is all about to change. From now on, all these meters will be replaced with wireless meters. This means that no one will have to come around and take a reading any more. The wireless equipment attached to the meter will automatically send a reading to the city each month."

"Since the wireless meter is installed on our property, we will be able to access the meter reading any time we want to. The city provides the software that allows us to access the meter readings. This software does the calculations and shows us how much we owe at any point in the month. We then have the ability to get our power and water bill information whenever we want, not just at the end of the month. Though, of course we still pay at a given date each month. I'm thinking that this may help us to be more responsible with using water and power. Perhaps as a community we will stop wasting those precious resources. I mean, personally, if I see that certain activities suck up a lot of power, I might limit those activities. I'd be curious to see how the bill changes if we unplug all of our appliances when they are not in use, or turn the water off as we are brushing our teeth."

Jerry was listening to this and getting just as excited as Julia. He decided to see how this new system could best be applied at his work. His company in the Silicon Valley had offices in ten buildings and each building had four floors. Jerry wrote a proposal regarding how to implement wireless utilities readings in each of the buildings and then offered this proposal to the people in charge at his company. They read the proposal and enthusiastically decided to implement this new system.

Before long, the company had installed these wireless power and water meters in each floor of every building. They also built an internal Website that captured in real time the power and water consumption of their company as a whole, and also for each building. The CEO sent out an email to all the employees in the Silicon Valley about the internal Website and asked for their cooperation in improving water and power consumption.

When the employees looked at the internal water and power consumption Website, they were alarmed at the amount of power they were consuming. All the power and water readings were stored as historical data and available at any point to any interested person, and for the creation of graphs.

The employees were interested in lessening their carbon footprint, and began coming up with new ideas for doing so. They started powering off the video monitors when they left for home in the evening. Immediately, they began seeing improvements in the rate of energy consumption. Now they turned off the monitors even when they went out for lunch. Some employees started powering off the entire system when they left for home. It took five to ten minutes to boot the system in the morning, but they were OK with that, as they could see what a difference it made. It made everyone feel as though they were doing something real and tangible to help the planet.

When information like this is available on a real-time basis, it will most likely lead people to act with greater responsibility when it comes to energy and water conservation. Jerry was watching the local news on TV one night when he heard that the city was promoting the installation of solar panels on houses. Jerry already knew that the federal government was offering a tax credit for solar additions, and he was interested in learning more. On the news, the anchor said that in order to promote solar panels, city officials had convinced citizens who had already installed solar panels on their houses to install wireless energy meters as well. This data was then made available to the general public.

So on a real-time basis, the city Website was collecting information from these wireless meters regarding the quantity of power generated from solar energy. Anyone could log on to the site and view the statistics, comparing the data with that from their own home. Jerry immediately logged on to the site and saw how drastically different their own bill was from those of the solar powered homes. This was all Jerry needed to go ahead and move forward with solar installation. Jerry talked with Julia about it and together they decided to go for it.

Jerry wanted to explore more about solar and wind energy. After doing some research, he identified a remote location two hours' drive from the Bay Area where it was sunny and windy most days of the year. Jerry decided to buy a small farmhouse in that tiny town. Compared to the Bay Area, the land was extremely cheap. There he installed a few solar panels and wind turbines on a trial basis. He also installed the wireless energy meter.

From home, he could now see how much energy was produced from the solar and wind power at his farmhouse a couple of hours away. After watching the data for a few months, he was absolutely convinced that it was a good investment. He talked to the power company and asked them if he could sell back the power generated from his farmhouse. As an unexpected bonus, the power company said that they already had the necessary infrastructure to take power generated from Jerry's farmhouse and send it back to his home in the Bay Area.

The solar power generated from the farmhouse turned out to be income-generating. Initially it was matching his consumption at the Bay Area house, but after a while, he began to see a surplus of energy in his power account.

NATURE

In this story, we see how Web on-the-go technology helps combat global warming and other ecological problems.

Summer vacation had rolled around again. The family piled in the car and headed to Yosemite National Park. They decided they would camp at the campground, and so they packed their tents and sleeping bags. Late one evening, the family found themselves sitting around their campfire roasting marshmallows for s'mores, and talking. Each person shared their ideas on using Web on-the-go technology for topics relating to nature. Nancy told the family what she had discovered on her MID about tracking birds and whales. Jonathan shared what he had found about real-time information on fish harvesting. Jerry told the family how MIDs could be used as a large natural disaster warning system (for example, to warn of a tsunami), and Julia spoke of the issue that was most important to her: global warming. After finishing off two s'mores, Nancy began to talk.

"We have been learning about animal migration patterns in school," Nancy began, "and what interested me most was how we have learned to track the migration of birds."

The family settled in to listen. Jerry and Julia were so proud of Nancy and Jonathan for their curiosity and interest in new subjects.

"We've all heard of the Audubon Society," continued Nancy, "but I had never really known the history of this organization till we learned about it in school. So, in 1803, a man named John James Audubon, a naturalist from America, began wondering about bird migration. He was curious as to whether or not a single bird would return to the same place each year. He decided to figure it out, by tying a small string around a bird's leg before it flew south. The following spring, Audubon saw the same bird with a string around its leg in the same place he had seen it before. So he knew that birds returned to the same place each year."

"But the cool thing is," Nancy continued, "scientists have found a much more effective way of tracking animals. They tag them with small tags that are electronic and give off repeating signals that are picked up by satellite radio devices. Obviously, this means that the scientists have a steady stream of data coming in regarding a tagged animal, without ever having to catch the same animal twice. But these electronic tags come with limitations, too. They are really expensive, and a bit heavier than a non-electronic tag, which some people think might slow birds down."

"I had no idea they were doing that," said Julia. "Go on."

"So," continued Nancy, "two separate devices are needed to track an animal: a transmitter, attached to the animal, sends out radio waves just like any radio station. Then a receiver is needed to pick up the signal. This receiver is usually in a truck or an airplane, because scientists must follow the animal in order to keep track of the signal. The receiver can also be put in satellites that are orbiting the Earth. Scientists use networks of these satellites to track animals. The signals from all the satellites work together to figure out exactly where an animal might be. They can also watch the animal move and thus see its path. This type of tracking, using satellites instead of trucks, is really great because people don't have to follow the animals wherever they go. Using satellites, scientists have tracked migration patterns of caribou, sea turtles, whales, seals, elephants, bald eagles, and ospreys."

"Since they've begun using this electronic tracking system, scientists have gained a much more accurate picture of migration patterns. One interesting example of this is when scientists electronically tracked a herd of caribou. In doing so, they learned that the herd moves way more than they had thought, and that they return each year to the exact same spot to give birth. Scientists couldn't have figured this stuff out without the electronic tags."

"Another cool example of how important this tracking system is involves manatees. Manatees are an endangered species and scientists are trying really hard to protect them. Using radio tracking, we've learned that manatees have traveled as far north as Rhode Island during their migration, which means we must protect the entire Eastern seaboard in order to protect them, whereas before, protection efforts were limited to the Florida coast. So this tool can be really helpful in helping endangered species."

"Wow," said Jerry, "that's so cool!"

"There's more," said Nancy. "The same technology that we are using to track animals could really help the people whose work or recreation somehow is affecting these animals. The example we learned about in school involved salmon. A little further north than where we live on the Peninsula, there are protected areas where the salmon run up the creeks. Over the years, park officials have noticed a decline in the numbers of salmon, so they decided to ban any fishing in any waterway in the park throughout the winter and spring when the salmon are swimming upstream to spawn, and any boating on the rivers. However, if they had more detailed migration information, officials might have more options. It could be that they can decrease the amount of time the ban is in effect, or limit the bans to the exact places in the rivers and streams where the salmon run, instead of the whole park."

"But the coolest thing," continued Nancy, "is that even this electronic tagging system is getting much better these days. Now, with Web on-the-go technology, tracking birds, whales, and other animals is way easier. Instead of heavy electronic tags that might slow small animals down, scientists are now able to attach a mini-MID that is lightweight and solar-powered. They are able to cut down on many of the features from the MID, such as the screen, keyboard, and camera, and build a mini-MID."

"So now scientists can program these devices with an animal identification code, and the frequency with which data should be sent to the Website and address of the web server. The device then sends the animal identification code, current GPS location, and the current date and time to the Web sites at the programmed intervals. The migration data is captured and stored on the Website automatically. There is no need for either manual or satellite tracking. The mini-MID attached to the animal reports its location to the Websites automatically. This frees the scientists up to analyze this data and work on the more important aspects of species conservation."

"Before this technology, scientists needed to follow the tagged animal in order to collect current data and often simply lost track of them. With the new Web on-the-go technology, the device sends the data automatically and tracking has become much easier and much more convenient. Now scientists can work comfortably with this new information and predict the animals' behavior much more accurately."

"And lastly," said Nancy, "we can use this device for domestic animals, too. People can attach this mini-MID to their pet's collar, and track it using this device. There would really be no way for the pet to be lost ever again."

"Nice," said Jonathan. "Maybe we should get something like that for Snowflake."

"Well," said Jonathan, "I have a similar kind of story. In biology class we've been learning about how the fish populations in the oceans are diminishing. They say that 90 percent of the big predator fish are now gone from the oceans, due to overfishing. Isn't that scary?"

Everyone nodded their heads.

"We learned that fish make up about seven percent of the world's total food supply, and in developing countries, this number is much greater. Our teacher said that nearly half a billion people earn their livelihoods from fishing or harvesting the oceans. You can imagine, however, that these people are having a very tough time locating and catching the fish as fish stocks dwindle and move further offshore. This makes the job of a fisherman difficult, as he has to travel further and come home with less."

"So there are many things that must be addressed," continued Jonathan. "We need to make sure that we don't completely overfish the oceans, while at the same time, make sure that people are still being fed and able to earn a living."

Many countries have decided that it is important to identify the places where fish actually are, in order to locate fish stocks and allow fishermen to catch more. India is a good example of this.

Over the past ten years, India has developed a system of helping fishermen locate fish. Using oceanic features such as temperature fronts, meanders, eddies, rings, and upwelling areas, they have been able to locate the sites where fish will most likely congregate; then these places are identified using satellite imagery. This has helped the fishermen out a great deal.

These oceanographic features can be mapped in near-real time and are used to locate potential fishing zones (PFZ) that are then made available to the Indian fishing community, which consists of nearly six million fishermen. The Indian National Center for Ocean Information Services (INCOIS) gives the PFZ advisories in local languages, three times a week, to the entire coast of India via fax, phone, Internet, email, newspaper, and radio. These advisories let people know where the fish are likely to be in the next two to four days, and offer detailed directions on how to get there. Using these advisories, fishermen have reduced the time it takes them to find fish by 70 percent and have increased their catch.

"This is a great example of how science can help the common man," continued Jonathan. "But I've thought about this a great deal, and there seem to be some limitations to this system as well. With the current scenario, when the organization receives the information about the PFZ, processes this information, and then sends it out to the general public, by that time the data will already be two days old. Hopefully the fish stay at the same location, but they very well might not. It seems to me that it is so important that the information is delivered in real time and with extremely precise locations."

"And I came up with an idea for how to make it better," said Jonathan, "using Web on-the-go technology. What if every fisherman could carry a Mobile Internet Device? Using that device he could subscribe to the notifications about the fish stocks in the ocean. Once organizations have identified the PFZ, they immediately update their website with that information. They can even provide the GPS location of those potential fishing zones. All subscribed persons will get an alert and they will head out into the ocean with their MID in hand. Now that they know the GPS location, the MID in fact provides driving directions so the boats can reach their destinations."

"Well, that sounds good for the fishermen today," said Nancy, "But what about ten years from now? I mean, the real problem is that there aren't enough fish, and this affects people. If we keep allowing people to fish just as hard as they have always been fishing, then the problem will just keep getting worse! What you're talking about seems like a very short-term solution."

"Well, I'm not done yet," said Jonathan, "and I agree with everything you've said. So what they are doing now, is using this exact same technology to find areas where the fish often return over time. While some of these areas will remain open for fishing, some of these zones will be declared 'no-fishing zones' and anyone caught fishing in them will be heavily fined. No-fishing zones will be kind of like preserved habitat areas where humans are no longer allowed. In this way, fish stocks will have safe areas in which they can recover. Officials can use the same method of disseminating information to fishermen, to let them know what zones are off limits.

If all fishermen have access to the PFZs, they will also have access to the no-fishing zones, and will have no excuses if they are found fishing in these zones. Indeed, establishing no-fishing zones might be hard for some fishermen in the meantime. But there are other solutions as well, like helping these people to become fish farmers instead of harvesters, or introducing other, easily-farmed meats into the local diet."

"These sure are interesting times," said Jerry.

"Well," said Jerry, "thank you both for letting us in on such important information. I'm so glad that you guys are interested in helping to preserve life on this planet. I have been thinking about another way in which Web on-the-go technology could be really helpful for humans and the Earth. Do you all remember the 2004 tsunami?"

Everyone nodded their heads.

"Amazingly, the earthquake generated by this tsunami is estimated to have released the same amount of energy as 23,000 Hiroshima-type bombs going off at once, at least according to the U.S. Geological Survey."

"Do we have any idea why it happened?" asked Julia

"Well, I've read that tremendous forces had been building up deep inside the Earth for hundreds of years. On that day, December 26, 2004, these forces were suddenly released, causing the ground to shake violently. A series of huge killer waves were released and moved across the Indian Ocean as fast as a jet airliner."

"Yikes," said Nancy.

"So, National Oceanic and Atmospheric Administration (NOAA) scientists at the Pacific Tsunami Warning Center in Hawaii went to work immediately after they received a seismic signal that an earthquake had occurred. They issued a bulletin to Hawaii, the West Coast of North America, and many other countries that there was no threat of a tsunami to their coastlines. And then they went to work notifying other countries about the potential of a tsunami after the earthquake that had registered as a 9.0."

"However, the Pacific Tsunami Warning Center did not detect a tsunami in the Indian Ocean, as there were no buoys in place there. They had no idea that they should be contacting the authorities in the Indian Ocean region, and meanwhile the tsunami was traveling at about 500 miles per hour."

"By the end of that day, more than 150,000 people were dead or missing and millions more were homeless. The devastation ranged across eleven countries, and many considered it the most devastating tsunami in history."

"So the Pacific Tsunami Warning Center in Hawaii had detected the earthquake, but there was no tsunami alert system in place in the Indian Ocean."

"Some countries, including Thailand, criticized NOAA, saying they should have done more to raise the alarm. In response to this criticism, NOAA officials said that there was no proper system in place for these countries to receive a tsunami warning. After the tsunami, plans were drawn to expand tsunami detection and warning capabilities."

"Implementation of this new plan would include enhanced monitoring, detection, warning, and communication of such disasters around the globe."

"So in response to the disaster, NOAA plans to deploy thirtytwo new deep-ocean assessment and reporting of tsunami (DART) buoys, including areas throughout the Pacific and Caribbean basins. These buoys are able to record surface heights of the ocean via satellites."

"A tsunami warning system (TWS) is a system meant to detect tsunamis and also to issue warnings to populations in order to prevent loss of lives and property. A TWS includes two equally important components: a network of sensors to detect tsunamis and a communications infrastructure that is capable of delivering rapid alarms so that evacuations can take place. The DART system is a part of this enhanced tsunami warning system."

"Each DART station consists of a sea floor bottom pressure recording (BPR) package that picks up on pressure changes caused by tsunamis, and a surface buoy. The surface buoy is able to receive information that has been transmitted from the BPR and then transmits data to a satellite. At this point, the satellite retransmits the data to ground stations and the information is immediately disseminated to NOAA's Tsunami Warning Centers. By using this system, NOAA is able to send warnings to all areas that might be in danger."

"As the system stands now, the DART station communicates to the NOAA warning centers via satellites. Following this, the scientists process the data manually and then issue warnings," Nancy concluded.

"Wow, that's a little bit complicated," said Julia. "Is there any way to simplify that process?"

"Well," replied Jerry, "I do think there is a way to implement the Web on-the-go technology in this instance as well, which would simplify things a great deal. Imagine that each DART station had an MID installed. The MID could periodically send the change of pressure data along with the GPS location to the appropriate Websites via the local wireless infrastructure instead of the satellite system. Then, on the Web, the proper software could be available that would process the incoming data and issue warnings to all those who were subscribed to the program. The entire process could be automated and scientists could simply monitor it and act appropriately if there happened to be an exception."

"Since the satellite link for communication is eliminated, the DART station would be much cheaper and thus it would be possible to install more DART stations in the ocean. Since all data would be collected and stored on the Web, the data would be readily available for further processing and also to retrieve historical data when needed to make predictions about future natural disasters."

"And because this information is stored on the Web, any authorities around the globe who are interested in these alerts can subscribe to the program."

"The MID could be used not only for the tsunami warning system; also to monitor and report other natural disasters such as hurricanes, tornadoes, and floods."

"Wow. That's quite interesting. Thank you all for sharing such valuable information with me," said Julia. "All this talk of natural disasters makes me think of global warming. I always wondered why it took such a long time to bring global warming to the attention of the people. I mean, this is an issue that has probably been happening for quite some time. But only recently have people really become concerned about it. If you ask me, this is a huge issue! I think the main reason for this delay is simply due to a lack of information on a real-time basis."

"Until recently, there was no equipment or system available to monitor the climate changes on Earth. People might have wondered about how their pollution affected the air and climate, but tracking change was virtually impossible."

"I'm not sure if you all are aware that many countries have what we could call an 'environmental antennae.' This antenna is made up of orbiting satellites as well as other sensor equipment, such as the tsunami warning system, meant to monitor specific Earth details. Recently, the world's countries met with the goal of finalizing plans to link up all of these antennas, and thus form a sort of international fleet of observation equipment."

"Right now these systems are being used mainly to estimate crop yields, detect earthquakes, forecast droughts, predict floods, and monitor air and water quality."

"By linking the satellites and sensors together, the sensors will be able to 'talk to each other,' allowing information to be shared across continents. This would be a great asset in alerting nations to the threat of looming natural disasters."

"The nations decided to call this system the Global Earth Observation System of Systems (GEOSS). The program hopes to pool all national and regional observation data within a decade. Once this system is in place, information would be available immediately to each and every subscribing country."

"But if we instead used the Web on-the-go technology, these sensors wouldn't need to talk to the orbiting satellite. Instead, they could simply locally access the wireless infrastructure."

"The sensors could be attached to MIDs. The MID could then send the date and time, GPS information, current temperature on to the appropriate Website. There would be no need for manual data entry."

"And, as you mentioned, Jerry, Web on-the-go technology is so much cheaper and establishes a system of two-way communication. Right now most satellites have their own unique purpose, like tsunami detector satellites or weather detector satellites. Using Web on-the-go technology, however, the satellite would have a more general purpose. Satellites would provide wireless access so that these other systems are able to function. More than one system could be using one satellite at the same time."

"Hey," said Jerry, "I can already envision how efficient that would be. I mean, if we are able to get a thorough picture of what is actually happening on the earth in terms of climate change, natural disasters, ice levels, and so forth, we would be much better positioned to make the right decisions to combat global warming."

"Exactly," said Julia, "That's why it's such an exciting idea."

THE BEGINNING

This is not the end. Rather, it is a beginning. In the initial chapters of this book we have laid the groundwork for the Web on-the-go concept. With the help of ten fictionalized examples, we illustrated potential Web on-the-go applications. By now it is understood that wireless infrastructure is a key element for the success of this concept. It is of the utmost importance that city governments, county governments, state governments and the federal government work quickly to make policy decisions that will provide the public sphere with a widely accessible wireless infrastructure.

We can think about how to begin making appropriate policy decisions at these different levels of government by using the example of transportation infrastructure. We drive our cars in the city on roads that are maintained by city governments. Freeways and expressways, however, are maintained by county governments. Highways are maintained by the state governments, whereas interstate highways are maintained by the federal government. In addition, we have seaways for ships and airways for airplanes. It is the responsibilities of various government organizations to provide infrastructure that meets the transportation needs of the people as they move about their lives. Similarly, it is the responsibility of various government organizations to provide wireless infrastructure to meet the communication needs of our global society.

In democracy, it is said that we have a "government by the people, government of the people and government for the people." However, more often than not, people look at the government as, of the politicians, by the politicians, for the politicians. When we look at our political leaders as politicians, we don't consider them as part of "the people." Many of us have learned to view our politicians as if they were aliens from some other planet, but really they are also included in "the people." Instead of "politicians," it might be more helpful to view them as policy makers. It could be easier to see them as part of "the people." Policy makers represent the people in the Council, Senate and Congress. They make important decisions that change the world and transform it into a better place to live.

One relevant example, and one worth investigating here, is how decades of policy makers have all made very important decisions regarding GPS (Global Positioning System) technology. It was these decisions that brought GPS technology to the wider public. Had these policy decisions not been made, GPS might not be the service that is today, available free to anyone on the planet. This is a technology that we can thank every day for landing thousands of commercial jets and boats safely, and providing important data for emergency rescues and wildlife monitoring.

GPS is a U.S., space-based radio navigation system that provides accurate positioning, navigation, and timing services to ordinary citizens 24 hours a day, all around the world, at no cost. For anyone with a GPS receiver, the system will always provide a current location and time for an unlimited number of people, no matter the weather, day and night, anywhere in the world. This is incredible, if you think about it, and not to be taken for granted. And really there's no reason why wireless access shouldn't be the same.

The GPS is made up of three separate parts: GPS satellites revolving around the Earth, control and monitoring stations on the surface of the Earth, and GPS receivers owned by individual users. The satellites broadcast signals from space that are then picked up by GPS receivers.

Anyone can purchase a GPS handset from almost any electronic store. Once someone has a GPS device, it can be used to determine one's present location, destination, and the time, whether the user is walking, driving, flying or boating. Of course, there are many layers of information that can then be accessed including elevation gain and loss, rate of travel, etc. Transportation systems all around the world have begun relying on GPS to provide navigation for aviation, ground, and maritime operations. GPS has become a priceless tool for disaster relief organizations, as well as for geologists, farmers, surveyors and innumerable other types of users who rely on the service for their livelihood and also for recreation.

Three American presidents, President Ronald Reagan, President Bill Clinton and President George W. Bush, all made very important policy decisions regarding GPS. These decisions in turn have become a major factor that led to the success and growth of GPS applications all around the world. With these kinds of policy decisions affecting ordinary people all over the world, the United States has demonstrated that it is not solely concerned with being the SuperPower anymore, but rather that it is willing to be Super Responsible for the global community. The decisions in this arena made by the American presidents have had a huge impact on the lives of the people living all over planet Earth.

Many people don't realize that for over twenty years, the United States government has adopted policies that encourage the worldwide use of GPS and other space-based PNT (Positioning, Navigation, and Timing) services. Because the U.S. policies around GPS are stable and transparent, and GPS service has been so dependable, global innovation and competition around GPS technology have sped up, leading to the development of new industries and applications based on GPS.

The consistent U.S. policy to provide open access to civil GPS signals, free of charge, began in 1983, directly following the loss of Korean Airlines Flight 007 in Soviet airspace. Since that time, the U.S. government has made sure to continue this policy through presidential directives, congressional law, and international agreements.

Korean Air Lines Flight 007 was a Korean Air Lines civilian airliner that was shot down by Soviet jet interceptors on September 1, 1983, over the Sea of Japan, just west of Sakhalin Island, when it was spotted flying over prohibited airspace. 269 passengers and crew were killed. Lawrence McDonald, a sitting member of the U.S. Congress, perished in this incident. The airplane was heading from New York to Seoul when it strayed accidentally into prohibited Soviet airspace due to a mistake in navigation. Only sixteen days later, President Reagan offered to make GPS available to be used by civilian airplanes, free of charge, when the system became operational. This marked the beginning of the spread of GPS technology from the military to civilian airplanes.

President Reagan also announced at that time, that the Global Positioning System would be made available for civilian use around the world as well (not just in airplanes), once the system

became operational. This was followed by a U.S. offer to make the Standard Positioning Service of GPS available, free of charge, starting in 1993, continuously, and for citizens all around the globe.

In 1995, President Clinton agreed with the government's earlier commitment to provide GPS signals to international users. A year later, he created the Presidential Decision Directive (PDD) on GPS, which maintains a clear U.S. commitment to the principles of open access and no charges for use of the GPS system.

In 2004 George W. Bush gave us what is now the primary document concerning GPS access and related policy issues. This document is called the U.S. Space-Based Positioning, Navigation, and Timing (PNT) Policy, and basically expands the scope of earlier directives, encouraging commercial use of GPS by ensuring that the civil signals will remain free of charge, as well as the technical information needed to actually make use of the signals.

This policy indeed demonstrates real foresight on the part of the U. S. in meeting the needs of GPS users around the globe. Perhaps the real significance of this grand success story is that people all around the globe can be confident that GPS technology will continue to evolve and meet the needs of ordinary citizens and commercial users alike. And thus new applications and markets for GPS technology are springing up every day.

Back in 1983 when President Ronald Reagan made the policy decision to make the GPS available for civilian use, he probably did not foresee the current growth of applications based on GPS. We are in a similar place right now when it comes to wireless infrastructure. Policy makers at all levels are faced with important decisions regarding wireless infrastructure in our society. And there is absolutely no doubt that once reliable wireless infrastructure is made available to the people, there will be a tremendous growth of applications available to use it.

Let's return again to the example of transportation infrastructure: In order to meet the transportation needs of the people, local and federal governments must build and maintain functional roads. Once the roads are built, people may make use of them in whatever vehicles they wish: cars, vans, trucks, sports cars, luxury cars, school buses, motorcycles, and bicycles. Oftentimes, one person has multiple vehicles, for different uses. This person may use a hybrid car to commute from home to work, while he may use his sports car for weekend trips, and his motorcycle for a sense of adventure. These different vehicles are most likely manufactured by different companies.

Likewise, to meet the communications needs of the people, local and federal governments must build and maintain reliable, consistent wireless infrastructure. Of course, then people may use different devices to access the wireless infrastructure. These devices will probably be manufactured by a variety of different companies.

There may be devices customized for specific applications. Amazon's Kindle is one such device meant for reading books, magazines etc. The Kindle was not popular until quite recently. When queried, people maintained that they were comfortable reading printed books, and would be less comfortable using a device that allowed them to read electronically. However, once Amazon made the Kindle with a wireless feature, then the perspective of many people instantly changed. With the ability to have an entire catalog of books available at their fingertips, this little device began to look a bit more exciting.

Digital content delivery is a green solution. With digital content, one really can save paper and trees, ink and chemicals. Also, the costs and ecological footprint involved in shipping and handling are totally eliminated.

Continuing with this digital content concept, I propose that there will be a new industry known as the Digital Content Server. May I suggest that we call it the Kindle Server? The Kindle Server could be the ideal product for libraries. In the future, libraries may have Digital Content Servers with loaded digital books, magazines and newspapers. Libraries may provide Kindle-like devices to their customers, or people may bring their wireless Kindle devices to the library. Using the Kindle reader, anyone can browse the catalog, select the book, and then read it, all using the Kindle Server.

This library Digital Content Server concept will be extended to airlines, trains and ships. On international flights, customers will have access to Digital Content Servers. These servers will have newspapers, magazines and books of all sorts available in the digital format. Once on the flight, a wireless Kindle reader will be provided to all the customers. Wireless Kindle is basically a Mobile Internet Device, customized for reading books.

Similarly, if you add wireless features to digital cameras and video recorders, people will undoubtedly come up with tons of innovative applications for the new technology.

So as you can see, as we reach the end of this book, really we are just getting started. Here we have only but scratched the surface. We have not explored potential applications in many industries (such as healthcare) that also will be making use of this new technology.

In future books, we shall continue to explore different industries, and then identify the potential for applications that will add enormous value to the lives of all people on planet Earth. As we continue to identify different industries that can make use of this new wireless infrastructure, we will build interesting stories around these industries and present these stories to our readers in an engaging way. These stories will be full of real ideas that can be used to pull our society out of recession and carry our communications industry into the future. After reading these stories, it will be up to the people to put these ideas into action, at all levels of our society.

Our mission is to continue the search for innovative opportunities and applications, based on the Web on-the-go concept, which will add immeasurable value to the lives of all people on Planet Earth.

REFERENCES

"People only see what they are prepared to see," by Ralph Waldo Emerson. (1803 - 1882)

Leonardo da Vinci.

Science fiction author Arthur C. Clarke.

The 1945 Proposal by Arthur C. Clarke for Geostationary Satellite Communications.

First Lunar Landing.

History of Transistors.

"A new generation of innovation is about to change the way technology interacts with people ... In the next few years we are going to take a leap into uncharted territory."

Steve Ballmer, Microsoft CEO (February 17, 2009)

Intel is a trademark of Intel Corporation. www.intel.com

Google, Android, YouTube, Latitude are trademarks of Google Inc.

Apple, iPhone are trademarks of Apple Inc.

Microsoft is a trademark of Microsoft Corporation.

511.org. In 2001, the Federal Communications Commission designated 511 as the single travel information phone number for states and local jurisdictions across the country.

James Kim's story

OnStar is a trademark of General Motors.

Costco Wholesale Corporation is the largest membership warehouse club chain in the world.

Home Depot is an American retailer of home improvement and construction products and services.

www.ingramcontent.com/pod-product-compliance
Lightning Source LLC
Chambersburg PA
CBHW031857200326
41597CB00012B/447